Design, Analysis, and Measurement of Antennas

Design, Analysis, and Measurement of Antennas

Editor

Naser Ojaroudi Parchin

MDPI

Basel • Beijing • Wuhan • Barcelona • Belgrade • Novi Sad • Cluj • Manchester

Editor
Naser Ojaroudi Parchin
Edinburgh Napier University
Edinburgh
UK

Editorial Office
MDPI
St. Alban-Anlage 66
4052 Basel, Switzerland

This is a reprint of articles from the Special Issue published online in the open access journal *Applied Sciences* (ISSN 2076-3417) (available at: https://www.mdpi.com/journal/applsci/special_issues/design_analysis_measurement_antennas).

For citation purposes, cite each article independently as indicated on the article page online and as indicated below:

Lastname, A.A.; Lastname, B.B. Article Title. *Journal Name* **Year**, *Volume Number*, Page Range.

ISBN 978-3-7258-0059-9 (Hbk)
ISBN 978-3-7258-0060-5 (PDF)
doi.org/10.3390/books978-3-7258-0060-5

Contents

About the Editor

Naser Ojaroudi Parchin

Naser Ojaroudi Parchin is currently an assistant professor (lecturer) at Edinburgh Napier University, UK. He obtained his Ph.D. in Electrical Engineering from the University of Bradford, UK, where he was a postdoctoral research assistant at the Faculty of Engineering and Informatics and worked as a research fellow in the SATNEX V project, funded by the European Space Agency. From 2018 to 2020, he was a Marie Curie Research Fellow in the H2020-ITN-SECRET project funded by the EU Commission, targeting 5G mobile small cells, spread across four leading universities/research institutions and four industrial partners in five different European countries. His research interests include multi-band/UWB antennas, phased arrays, MIMO systems, smartphone antennas, SAR/user-impact, full-duplex diversity antennas, 5G antennas, implementable/biomedical sensors, RFID tag antennas, millimeter-wave/terahertz components, fractal structures, metamaterials and metasurfaces, Fabry resonators, EBG/FSS-Inspired radiators, band-pass/band-stop microwave filters, and reconfigurable structures. He has over 12 years of research experience in antenna and microwave engineering with good hands-on lab skills and experiences as a laboratory assistant and demonstrator. He also supervised several BSc/MSc students for their final projects and supported several Ph.D. students in their research. He is the author and co-author of several books/book chapters and more than 300 technical journal and conference papers with more than 7400 citations and an H-index of 48.

Preface

Dear Colleagues,

Rapid growth in the area of modern wireless communication systems has led to the demand for multifunctional devices to be used in various wireless services. Antennas for these devices must be compact, multifunctional, and able to maintain a high level of performance in any kind of environment. One of the important issues in communication systems is the provision of the right orientation between the transmitter and receiver antennas. Circular and dual polarized antennas are good solutions to this problem. Reconfigurable and switchable antennas are attractive in various applications such as cognitive radio, surveillance, and tracking because they produce more than one operation band and radiation pattern using active elements. MIMO technology and phased array systems with multiple adaptive and smart antennas can significantly enhance the capacity of the system and resist multipath fading, and are promising technologies with which to meet the requirements of the future 5G networks. In a MIMO antenna system with limited space, one of the urgent difficulties to be resolved is to reduce the mutual couplings from adjacent elements. The miniaturization of wireless and handheld devices using miniaturized antennas with metamaterials, EBG, and FSS is necessary nowadays. Low-profile antennas with a low-cost manufacturing and measurement process are desirable for various applications such as RFID, UWB, and WBAN systems.

The objective of this Special Issue is to shed some light on recent advances and novel approaches in the design, analysis, and measurement of antennas for various emerging wireless communication systems and to identify further avenues for the development of research and techniques in this exciting field.

Submissions can focus on conceptual and applied research in topics including but not limited to the following:

MIMO and array antennas;
UWB antennas;
Mutual coupling reduction;
Antenna optimization;
Electromagnetic bandgap (EBG) sructures;
Multiple 5G antennas;
Antenna miniaturization;
Mobile phone and handheld antennas;
Filtering antennas;
Circular and dual polarized antennas;
RFID antennas;
MM-wave and THz antennas;
Adaptive and smart antennas;
Metamaterial antennas;
Fractal antennas;
Antennas for biomedical and wireless body area networks;
Automotive, radar, and satellite antennas;

Reconfigurable and switchable antennas;
Prototyping and manufacturing methods;
Measurements and experimentation of antennas.

Naser Ojaroudi Parchin
Editor

applied sciences

MDPI

Editorial

Editorial on "Design, Analysis, and Measurement of Antennas"

Naser Ojaroudi Parchin

School of Computing, Engineering and the Built Environment, Edinburgh Napier University, Edinburgh EH10 5DT, UK; n.ojaroudiparchin@napier.ac.uk

The area of modern wireless communication systems has experienced rapid growth, leading to a rising demand for multifunctional devices capable of providing various wireless services. For these devices, the antennas need to possess key characteristics: compactness, multifunctionality, and consistent high performance across different environments. A critical aspect in communication systems is ensuring the correct orientation between transmitters and receiver systems [1]. Circular/dual-polarized antennas offer effective solutions to address this challenge. These antennas enable reliable and efficient signal transmission and reception. In addition, reconfigurable and switchable antennas have garnered significant interest in diverse applications, such as cognitive radio. By incorporating active elements, these antennas can operate across multiple bands and generate different radiation patterns, providing versatility and adaptability in their functions [2]. Employing MIMO and phased array systems with several smart antenna elements can significantly increase the system capacity. As a result, these technologies hold considerable promise for meeting the requirements of future 5G networks, where higher data rates and seamless connectivity are essential. However, implementing MIMO antenna systems in a limited space presents urgent challenges, particularly in reducing mutual couplings from adjacent elements. In today's context, miniaturization of wireless and handheld devices has become a necessity [3]. Miniaturized antennas with the integration of metamaterial cells, Electromagnetic Band Gap (EBG) structures, and Frequency Selective Surfaces (FSS) are gaining importance. These advanced materials and structures allow for efficient antenna designs, making them indispensable components in modern wireless technologies. Furthermore, the quest for small, low-cost antennas remains high on the agenda. These antennas are highly desirable for various applications, including Radio Frequency Identification (RFID) and Ultra-Wideband (UWB) systems [4]. As wireless communication continues to evolve, addressing these challenges and incorporating innovative antenna technologies will pave the way for more efficient, versatile, and reliable wireless devices and systems, meeting the ever-increasing demands of the modern world. Therefore, antennas play a pivotal role in modern wireless communication systems and, as technology continues to evolve, the demand for more efficient, compact, and multifunctional antennas becomes ever more apparent.

The scope of this Special Issue is to provide a comprehensive overview of the latest developments and innovative methodologies in the design, analysis, and measurement of antennas. It comprises a collection of 11 papers that explore various aspects of antenna design, analysis, and measurement for emerging wireless communication systems. Each paper delves into specific topics and advancements within the field, contributing to the overall objective of shedding light on novel approaches and encouraging further research into this exciting area. The brief explanations of these papers are as follows:

Yang et al. [5] introduce a low-profile Wi-Fi antenna designed to deliver optimized radiations for Uncrewed Aerial Vehicle (UAV) applications. Specifically engineered for mounting on small UAVs, on the non-metallic wing's outermost side, this antenna exhibits impressive impedance bandwidths across two frequency ranges: 2.11 to 2.58 GHz and 5.06 to 7.5 GHz. The antenna's radiation patterns have been carefully optimized at 5.8 and

Citation: Parchin, N.O. Editorial on "Design, Analysis, and Measurement of Antennas". *Appl. Sci.* **2023**, *13*, 10069. https://doi.org/10.3390/app131810069

Received: 1 September 2023
Accepted: 2 September 2023
Published: 6 September 2023

2.4 GHz, ensuring enhanced antenna gain as the small UAV operates at greater distances in urban and open environments, respectively. Moreover, it demonstrates excellent vertically polarized radiation, making it well-suited for long-distance communication applications.

Hussain et al. [6] investigate the development and verification of an antenna tailored for Ka-band 5G networks. The resulting antenna demonstrates remarkable performance across a broad frequency range, spanning from 26.5 GHz to 43.7 GHz. The proposed design incorporates stubs with loaded rectangular patches, strategically integrated to enhance impedance bandwidth and achieve ultra-wideband characteristics. To further enhance functionality, a MIMO antenna is created by combining multiple elements. The hardware prototypes of both the individual antenna element and its MIMO configuration are fabricated and tested. The measured results closely align with the simulated outcomes.

Rizvi et al. [7] present a new CPW-fed antenna specifically designed for UWB applications with a WLAN band notch. The antenna boasts a simple geometry, featuring a rectangular radiation element integrated with a Y-shaped resonator, which effectively enhances the impedance bandwidth over the frequency range of 3–14.55 GHz. To achieve rejection at 4.59–5.82 GHz, an additional stub is incorporated into the antenna's design. This feature enables the antenna to maintain constant radiations with a gain exceeding 2 dBi within the pass-band, while demonstrating a minimum gain of −3.9 dB in the notch band. The antenna's measurement results exhibit excellent agreement with the simulations, confirming the accuracy and effectiveness of the proposed design.

Fernandez et al. [8] propose a modified patch antenna with a notable design simplified by using air as the dielectric and a straightforward patch shape. Ensuring that the feeding pin rod connects to the patch to the microstrip line is critical, ensuring excellent impedance matching for maximum radiation at the resonance frequency. Operating at 2.4 GHz, the antenna achieves impressive performance metrics, including a 9.63 dBi gain and a 93.08% total efficiency. The antenna's advantages lie in its easy design and manufacturing process, absence of a dielectric, through-wire feeding technique, robustness, error tolerance, and the ability to achieve superior performance across all metrics simultaneously.

Hu et al. [9] analyze and discuss the fundamental properties, as well as signal propagation curves, of a single-tower umbrella antenna, which can be a suitable choice for enhancing Loran transmission. Detailed analysis and comparison of the radiation characteristics of the transmitting antenna along with simulations in the complex geographical environment of Tibet, China are discussed. The results demonstrate the advantages of using the single-tower antenna in Tibet, and led to the design of the transmitting antenna's structure and electrical parameters, resulting in the capability to extend the enhanced Loran signal transmission to 1000 km or even further.

Qashlan et al. [10] employ a microwave breast imaging model using nine identical Vivaldi antennas to detect cancer tumors in a multilayer model. The array's configuration involved one antenna as a transmitter, and the remaining eight as receivers, in order to measure the backscattering signal of the breast phantom. This process was repeated with each antenna taking turns as the transmitter. Various tumor sizes and locations were tested and, remarkably, the locations could be accurately determined regardless of the tumor size. The Vivaldi element used in the study features a compact size of 25×20 mm^2, and a unique geometry. This antenna was fabricated and demonstrated excellent agreement between its simulated and measured performance.

Ahmed et al. [11] propose a straightforward and efficient design for an ultra-wideband (UWB) antenna, which takes the form of a jug shape with a handle on the right side of the radiator. This UWB antenna is printed on a cost-effective FR-4 substrate, making it suitable for wireless communication systems. The antenna's dimensions are 25 mm $\times$ 22 mm $\times$ 1.6 mm. The performance of the antenna is impressive, with a maximum gain of 4.1 dB, covering the entire UWB spectrum from 3 GHz to 11 GHz. The antenna's performances are investigated through simulations and measurements. This design is particularly well-suited for wireless communication systems and portable devices, due to its compact size and excellent performance across the UWB frequency range.

Rahamim et al. [12] present an innovative concept for a tunable reflector metasurface (MS) that enables beam steering at 28 GHz. This has been achieved by using varactor diodes as the tunability component in unit cells of the metasurface. This new concept, called the "stripes configuration", leads to remarkable enhancements in beam steering capabilities. The results demonstrated a 3 dB improvement in reflectance gain, depending on the scanning angle, compared to a uniform metasurface with just one stripe. Moreover, the stripes configuration showcased an impressive enhancement of 50% in the steering accuracy of the angle for various frequencies.

Umair et al. [13] offer a high-gain and compact antenna that offers both low backscattering and improved gain characteristics. The antenna's construction involves a cavity formed by combining an absorptive Frequency Selective Surface (FSS) with a double-sided Partially Reflective Surface (PRS), placed above a patch resonator. This configuration achieves a wideband Radar Cross Section (RCS) reduction in incident waves, including in-band frequencies. The results demonstrate a wideband RCS reduction from 4 to 16 GHz, with an average RCS reduction of approximately 8.5 dB compared to a reference patch antenna. Additionally, the antenna achieves off-broadside peak radiation at an angle of -38°, with a gain approaching around 9.4 dB.

Niu et al. [14] explore a novel multi-node inverse Finite Element Method (iFEM) for sensing the shape of flexible structures using strain sensors. In accordance with the Mindlin plate theory, a weighted-least-squares function is minimized by considering all strain measures. This approach enables accurate approximation for large inverse finite elements, providing the advantage of simultaneous extrapolation and interpolation calculations for elements. By substantially expanding the size of elements, the proposed method decreases the number of sensors required while improving the accuracy of the reconstruction process. This innovation holds promise for enhancing shape-sensing capabilities in flexible structures, and can lead to more efficient and accurate strain measurement systems.

Liu et al. [15] introduce a novel method for generating multiple pseudo-Bessel beams, with precise control over their propagation directions using a reflective metasurface. In this technique, the reflective metasurface's miniaturized unit cell modulates the electromagnetic waves, generating pseudo-Bessel beams that are propagated in various directions off-axis. This metasurface is capable of generating dual pseudo-Bessel beams, and their propagation directions can be accurately controlled. The efficiency of these beams is measured at 59.2% when propagated at 400 mm. Overall, the method showcases the promising potential for producing high-efficiency multiple beams with precise control, which could find applications in various fields.

We would like to sincerely express our gratitude and appreciation to all the esteemed authors whose exceptional contributions have enriched this journal. Their valuable research has significantly advanced the field of design, analysis, and measurement of antennas. Our heartfelt thanks go out to the diligent reviewers whose insightful comments and feedback have played a crucial role in enhancing the quality of these articles. Their expertise and dedication have been instrumental in ensuring the rigor and accuracy of the published works. We also extend our acknowledgment to the dedicated editorial board and the supportive editorial office of MDPI's *Applied Sciences* journal. Their guidance and assistance throughout the publication process have been invaluable in bringing this collection to fruition. We are confident that the readers will find these articles informative and enlightening, offering valuable insights into the world of antenna design, analysis, and measurement. We look forward to continued collaboration and contributions from researchers in the field, as we strive to further advance and explore new frontiers in antenna technology. Thank you all for your remarkable efforts and dedication.

Conflicts of Interest: The author declares no conflict of interest.

References

1. Zhang, Z. *Antenna Design for Mobile Devices*; Wiley-IEEE Press: Hoboken, NJ, USA, 2017.
2. Ojaroudi Parchin, N.; See, C.H.; Abd-Alhameed, R.A. Editorial: A Planar Dual-Polarized Phased Array with Broad Bandwidth and Quasi-Endfire Radiation for 5G Mobile Handsets. *IEEE Trans. Antennas Propag.* **2021**, *69*, 6410–6419. [CrossRef]
3. Ojaroudi Parchin, N.; See, C.H.; Abd-Alhameed, R.A. Editorial: Special Issue "Antenna Design for 5G and Beyond". *Sensors* **2021**, *21*, 7745. [CrossRef] [PubMed]
4. Ojaroudi Parchin, N.; Jahanbakhsh Basherlou, H.; Al-Yasir, Y.I.A.; Abdulkhaleq, A.M.; Abd-Alhameed, R.A. Ultra-Wideband Diversity MIMO Antenna System for Future Mobile Handsets. *Sensors* **2020**, *20*, 2371. [CrossRef] [PubMed]
5. Yang, Y.-L.; Lin, D.-B. A Compact Planar Wi-Fi Antenna with Optimized Radiation Patterns for Small UAV Applications. *Appl. Sci.* **2023**, *13*, 7470. [CrossRef]
6. Hussain, S.A.; Taher, F.; Alzaidi, M.S.; Hussain, I.; Ghoniem, R.M.; Sree, M.F.A.; Lalbakhsh, A. Wideband, High-Gain, and Compact Four-Port MIMO Antenna for Future 5G Devices Operating over Ka-Band Spectrum. *Appl. Sci.* **2023**, *13*, 4380. [CrossRef]
7. Rizvi, S.N.R.; Awan, W.A.; Choi, D.; Hussain, N.; Park, S.G.; Kim, N. A Compact Size Antenna for Extended UWB with WLAN Notch Band Stub. *Appl. Sci.* **2023**, *13*, 4271. [CrossRef]
8. Fernandez, M.D.; Herraiz, D.; Herraiz, D.; Alomainy, A.; Belenguer, A. Design of a Wide-Bandwidth, High-Gain and Easy-to-Manufacture 2.4 GHz Floating Patch Antenna Fed with the Through-Wire Technique. *Appl. Sci.* **2022**, *12*, 12925. [CrossRef]
9. Hu, Z.; Dong, M.; Li, S.; Yang, C. Simulation-Based Selection of Transmitting Antenna Type for Enhanced Loran System in Selected Location. *Appl. Sci.* **2022**, *12*, 6031. [CrossRef]
10. Qashlan, A.M.; Aldhaheri, R.W.; Alharbi, K.H. A Modified Compact Flexible Vivaldi Antenna Array Design for Microwave Breast Cancer Detection. *Appl. Sci.* **2022**, *12*, 4908. [CrossRef]
11. Ahmad, S.; Ijaz, U.; Naseer, S.; Ghaffar, A.; Qasim, M.A.; Abrar, F.; Parchin, N.O.; See, C.H.; Abd-Alhameed, R. A Jug-Shaped CPW-Fed Ultra-Wideband Printed Monopole Antenna for Wireless Communications Networks. *Appl. Sci.* **2022**, *12*, 821. [CrossRef]
12. Rahamim, E.; Rotshild, D.; Abramovich, A. Performance Enhancement of Reconfigurable Metamaterial Reflector Antenna by Decreasing the Absorption of the Reflected Beam. *Appl. Sci.* **2021**, *11*, 8999. [CrossRef]
13. Umair, H.; Latef, T.B.A.; Yamada, Y.; Hassan, T.; Mahadi, W.N.L.B.W.; Othman, M.; Kamardin, K.; Hussein, M.I. Quarter Wavelength Fabry–Perot Cavity Antenna with Wideband Low Monostatic Radar Cross Section and Off-Broadside Peak Radiation. *Appl. Sci.* **2021**, *11*, 1053. [CrossRef]
14. Niu, S.; Li, K.; Liu, J.; Bao, H. A Refined Shape Sensing Method for Skin Antenna Structure Based on Inverse Finite Element Method. *Appl. Sci.* **2020**, *10*, 7620. [CrossRef]
15. Liu, H.; Xue, H.; Liu, Y.; Li, L. Generation of Multiple Pseudo Bessel Beams with Accurately Controllable Propagation Directions and High Efficiency Using a Reflective Metasurface. *Appl. Sci.* **2020**, *10*, 7219. [CrossRef]

applied sciences

MDPI

Article

A Compact Planar Wi-Fi Antenna with Optimized Radiation Patterns for Small UAV Applications

Ya-Lung Yang and Ding-Bing Lin *

Department of Electronic and Computer Engineering, National Taiwan University of Science and Technology, Taipei 106335, Taiwan; d10802204@mail.ntust.edu.tw
* Correspondence: dblin@mail.ntust.edu.tw

Abstract: This paper proposes a compact planar Wi-Fi antenna with optimized radiation patterns for small uncrewed aerial vehicle (UAV) applications in both urban and open areas. It is suitable for mounting on the outermost side of the non-metallic wing of small UAVs. It has small dimensions of 16.5 mm (L) by 30.3 mm (W) by 1.6 mm (h), and the measured results of its prototype are in agreement with simulated data. Its impedance bandwidths over the two frequency ranges are 2.11 to 2.58 GHz and 5.06 to 7.5 GHz ($|S_{11}| \leq -10$ dB). At 5.8 GHz, it has stronger radiation below the small UAV to reduce interference from rare-use directions. Its maximum radiations, the directions of the maximum radiation in each elevation plane, are below the UAV and between 14° and 29° from the horizontal plane. At 2.4 GHz, it has quasi-omnidirectional radiation to ensure a stable link in all directions, and its maximum radiations are near the horizontal plane. The optimized radiation patterns at 5.8 and 2.4 GHz can provide more antenna gain when the small UAV flies farther in urban and open areas, respectively. In addition, it has good vertically polarized radiation for long-distance applications.

Keywords: planar Wi-Fi antenna; optimized radiation patterns; small UAV; uncrewed aerial vehicle; vertically polarized radiation

Citation: Yang, Y.-L.; Lin, D.-B. A Compact Planar Wi-Fi Antenna with Optimized Radiation Patterns for Small UAV Applications. *Appl. Sci.* **2023**, *13*, 7470. https://doi.org/10.3390/app13137470

Academic Editor: Naser Ojaroudi Parchin

Received: 19 May 2023
Revised: 16 June 2023
Accepted: 22 June 2023
Published: 24 June 2023

1. Introduction

With the rapid increase in the number of small uncrewed aircraft vehicles (UAVs), their antenna design is becoming increasingly important to ensure a stable link [1]. This study addresses an antenna design for light-and-small civil UAV applications in urban areas. Small UAVs tend to be lightweight and small for portability, so they are unsuitable for mounting turntables or large phased-array antennas. Figure 1 shows the common scenarios of light-and-small civil UAV applications in urban areas where a person holding a remote control operates a quadcopter.

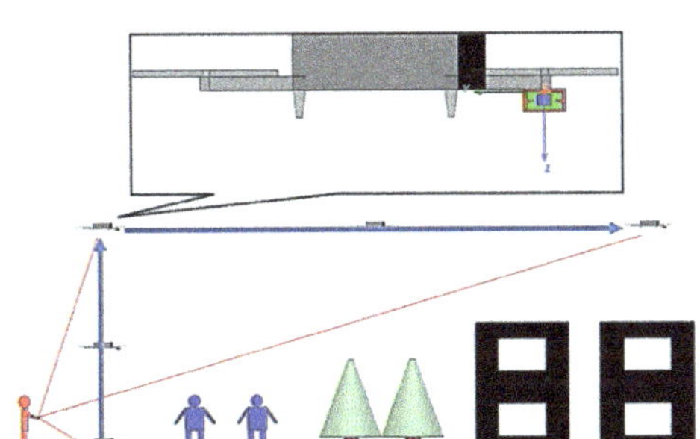

Figure 1. Common scenarios of light-and-small civil UAV applications in urban areas.

The proposed antenna is mounted below the propeller on the outermost side of the wing of a small UAV, away from large metal conductors, such as printed circuit boards

(PCBs) and batteries, in the UAV body. The UAV takes off from the ground, flies vertically above ground objects, and then flies forward to the farthest position. The flight must avoid hitting ground objects, such as people, trees, and buildings, so the UAV must stay above them. The distance between the UAV and the remote control increases as it takes off from the ground, flies vertically above ground objects, and flies forward to the farthest position.

To ensure stable transmission quality and stability, the gain and polarization characteristics of the antenna are determined based on the Friis transmission equation [2]. The antenna must meet the following requirements to reduce interference and provide more antenna gain when the small UAV flies farther in urban areas. First, its $|S_{11}|$ must be equal to or smaller than -10 dB to ensure efficient power delivery to the antenna. Second, it must have a small size and low profile to meet the portability and aerodynamic requirements for small UAV applications. Third, it must have stronger radiation below the UAV to reduce interference from rare-use directions. Given that the UAV is always above the remote control as it flies away, it is farther away from the remote control than when it was taking off. Fourth, its maximum radiations, the directions of the maximum radiation in each elevation plane, must be below the UAV and between 10° and 30° from the horizontal plane to provide more antenna gain when the small UAV flies farther in urban areas. Finally, due to the boundary condition of the earth's ground plane, the antenna must have good vertically polarized radiation for long-distance applications.

Although many antenna architectures have been proposed [3–14] to meet the requirements for UAV applications, none fully meet the requirements for small UAV applications in urban areas. Due to their small size and lightweight nature, folded dipole antennas have been proposed [3,4] for small and compact UAV applications, respectively. The fragmented antenna was proposed [3], and its overall dimensions and total mass are $0.96\lambda_0 \times 0.8\lambda_0 \times 0.8\lambda_0$ at 240 MHz and 18 g (including the matching circuit). A compact electrically tunable VHF antenna [4] was integrated into the landing gear of a compact UAV with good vertically polarized radiation. However, neither of them has stronger radiation below the UAV, and their maximum radiations are near the horizontal plane. In addition, both of them need an extra matching circuit to match their characteristic impedance to 50 Ω for a better reflection coefficient at the operation frequency. The low-profile, quasi-omnidirectional substrate-integrated waveguide (SIW) multi-horn antenna was proposed [5] for small non-metallic UAV applications. It has good vertically polarized radiation and a low profile of 0.028 λ_0 at 2.4 GHz. However, it does not have stronger radiation below the UAV, and its lateral dimensions of $0.76\lambda_0 \times 0.76\lambda_0$ are too large for small UAVs. Due to their low profile and ability to conform to planar and nonplanar surfaces, patch antennas were proposed [6,7] for UAV applications. The high-gain dual-mode cylindrical conformal rectangular patch antenna [6] has a low profile of 0.508 mm (0.0175 λ_0 at 10.323 GHz), stronger radiation below the quadcopter UAV, and good vertically polarized radiation. The required radius of the arm of the UAV is 25 mm at the lowest operating frequency of 10.288 GHz, and three patch antennas are needed on the UAV arm to cover the hemisphere below the UAV. The conformal patch antennas [7] have a low profile of 3 mm (0.058 λ_0 at 5.8 GHz) and stronger radiation below the UAV. However, their maximum radiations are directly above each patch. Both models proposed multiple patch antennas on the UAV to ensure a stable link in the required directions. However, it takes up too much surface area, making them unsuitable for small UAVs. The design of a combined printed helical spiral antenna and helical inverted F antenna [8] was miniaturized by co-winding the radiation elements on the same ceramic rod with a height of 24.7 mm (0.129 λ_0 at 1.57 GHz) and a radius of 6 mm. It was proposed for UAVs with a global positioning system (GPS) L1 band and telemetry communication frequency band (2.33 GHz) applications. However, it does not have stronger radiation below the UAV at 2.33 GHz. The low-profile broadband plasma antenna [9] has different radiation patterns at different operating frequencies and a low profile of 0.105 λ_0 at 30 MHz. Its radiation pattern is omnidirectional at 30 MHz and stronger radiation below the UAV at 300 and 500 MHz. However, it must be matched with extra Butterworth low-pass and high-pass filters to isolate the 13.56 MHz radio frequency signal to activate plasma. Its complex structure and the need to use special materials such as plasma will result in a heavier weight and higher cost, making it unsuitable for small UAVs. The broadband slotted blade dipole antenna [10] radiation patterns vary at different operating frequencies. However, it must be matched with a special matching circuit and does not have stronger radiation below the UAV. In addition, its radiation patterns at all operating frequencies have nulls near the horizontal plane, causing the smallest gain when the UAV flies to the farthest position. Due to the advantages of simple structure and low manufacturing cost using modern printed circuit technology, printed planar antennas were proposed [11–13] for UAV applications. The planar dual-mode dipole antenna [11] has a short electrical length of 0.3297 λ_0 at 0.86 GHz and good vertically polarized radiation. The omnidirectional vertically polarized antenna [12] also has good vertically polarized radiation and varying radiation patterns at different operating frequencies. However, neither has stronger radiation below the UAV, and their maximum radiations are near the horizontal plane. The wideband single-sided folded-off-center-fed dipole antenna [13] has good vertically polarized radiation and different radiation patterns at different operating frequencies. At 5 and 5.5 GHz, it has stronger radiation below the small UAV, and its maximum radiations are below the UAV and between 10° and 30° from the horizontal plane. At 3.5 GHz, it has quasi-omnidirectional radiation, and its maximum radiations are near the horizontal plane. However, its electrical length of 0.469 λ_0 is too long for small UAV applications. A compact slot coplanar waveguide-fed antenna [14] was not proposed for small UAV applications and has a very short electrical length of $0.16\lambda_0(L)$ and a low profile of $0.0128\lambda_0(h)$ to meet the portability and aerodynamic requirements for small UAV applications. However, it does not have stronger radiation below the UAV, and its maximum radiations are near the horizontal plane.

Summarizing the above literature review, most proposed antennas have maximum radiations near the horizontal plane for UAV applications. This feature can meet the requirements for UAV applications in open areas and provide more antenna gain when the small UAV flies farther. The wideband single-sided folded-off-center-fed dipole antenna [13] is the best; it has suitable radiation patterns to meet the requirements for small UAV applications in urban and open areas. In addition, it also has the advantage of a simple and cost-efficient structure, which can make light and small civil UAVs more affordable. However, its electrical length of 0.469 λ_0 at 3.5 GHz is too long to meet the portability and aerodynamic requirements for small UAV applications.

This paper proposes a dual-band antenna with optimized radiation patterns for small UAV applications in urban and open areas. It has a shorter electrical length than the best one proposed: 0.132 λ_0 at 2.4 GHz [13]. First, this study considers that the small UAV has a camera function and needs to transmit real-time images to the remote control for the user to preview or store. To meet this requirement, Wi-Fi 2.4 GHz and 5.8 GHz were selected as the operating frequencies. Secondly, since many technologies, such as Zigbee and Bluetooth, use 2.4 GHz as the operating frequency [15], there will be more interference at this frequency. Therefore, a quasi-directional radiation pattern was designed at 2.4 GHz, and its maximum radiations are near the horizontal plane for small UAV applications in open areas with less interference, such as mountains, oceans, or farmland. Third, an end-fire radiation pattern was designed at 5.8 GHz for small UAV applications in urban areas. It has stronger and maximum radiations below the UAV and between 14° and 29° from the horizontal plane. Finally, to ensure that the characteristics of this antenna will not be significantly affected when applied to small UAVs, the following assumptions are made in this study: First, as shown in Figure 1, the objects around the antenna, such as the UAV wing and propeller, must be made of non-metallic material. Second, the large metal conductors, such as PCBs and batteries, in the UAV body must be located in the far-field region of the proposed antenna. Based on the definition of the far-field region [16], the distance between the small UAV body and the proposed antenna must be > 46 mm.

By performing a theoretical analysis, full-wave simulation, and prototype measurement, we verified that the proposed antenna could meet the requirements for light and small civil UAV applications in urban and open areas. The rest of this paper is as follows: Section 2 presents the antenna design methodology, which includes the structural topology and theoretical analysis. Section 3 presents the parametric studies and full-wave simulations for validating the design theory. Section 4 presents the antenna prototype, measurement setup, comparisons of the simulated data, measured results of the prototype, and discussions about the results. Finally, Section 5 summarizes the results and future work.

2. Antenna Design and Analysis

In this study, a compact slot antenna with a coplanar waveguide (CPW) was designed that supports Wi-Fi dual-band communication and has a single-sided planar structure.

2.1. Antenna Structure and Main Design Steps

Figure 2 shows the structure of the proposed antenna with geometrical parameters. At 5.8 GHz, the four corners of its outer conductor are the primary radiation sources, forming end-fire radiation. At 2.4 GHz, the two sides of the outer conductor are the main effective radiation sources, forming quasi-omnidirectional radiation.

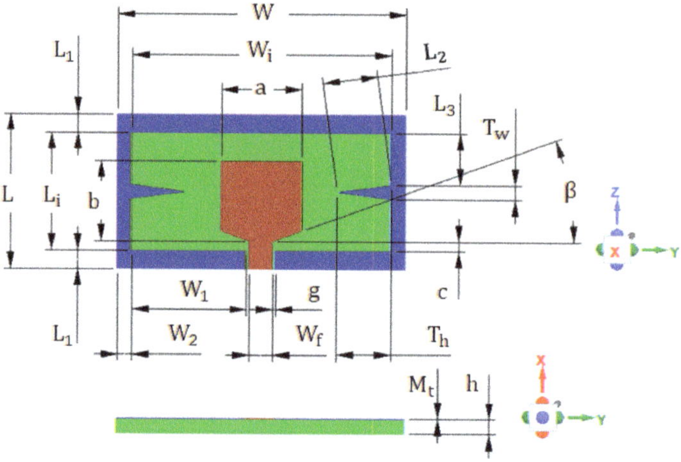

Figure 2. The structure of the proposed antenna with geometrical parameters.

Table 1 displays the values of geometrical parameters for the proposed antenna. The thickness of the metal layer (M_t) is 0.035 mm, of which 0.031 mm is copper, 0.003 mm is nickel, and 0.001 mm is gold. It was printed on the FR4 substrate with a relative permittivity of 4.4, a loss tangent of 0.02, a thickness of 1.6 mm (h), and planar dimensions of 16.5 mm (L) by 30.3 mm (W). At the fed port, the width of the central conducting track (W_f) was 2.5 mm, and a pair of outer conducting tracks were separated from the central conducting track by a small gap (g) of 0.3 mm to achieve a characteristic impedance of 50 Ω. The antenna was symmetrical on both sides of the x–z plane, so the radiation generated by the current components in the +y and −y directions would cancel out. Therefore, the currents in the +z and −z directions are the main radiation sources, providing the antenna with good vertically polarized radiation.

Figure 3 shows the main design steps of the proposed antenna. Steps 3 to 6 are the main design steps for key performances. The design theory is explained in detail in Sections 2.1 and 2.2.

Table 1. The values of geometrical parameters for the proposed antenna.

Parameter	Dimension (mm)	Ratio to the Wavelength of 2.4 GHz in Vacuum	Ratio to the Wavelength of 5.8 GHz in Vacuum	Angle (Degree)
W	30.3	0.24	0.59	N/A
W_i	27.3	0.22	0.53	N/A
L	16.5	0.13	0.32	N/A
L_i	12.5	0.10	0.24	N/A
W_1	12.1	0.10	0.23	N/A
W_2	1.5	0.01	0.03	N/A
L_1	2	0.02	0.04	N/A
L_2	5.75	0.05	0.11	N/A
L_3	5.5	0.04	0.11	N/A
a	8.5	0.07	0.16	N/A
b	8.5	0.07	0.16	N/A
c	1	0.01	0.02	N/A
T_h	5.7	0.05	0.11	N/A
T_w	1.5	0.01	0.03	N/A
W_f	2.5	0.02	0.05	N/A
g	0.3	0.00	0.01	N/A
C_t	0.035	0.00	0.00	N/A
h	1.6	0.01	0.03	N/A
Beta	N/A	N/A	N/A	20

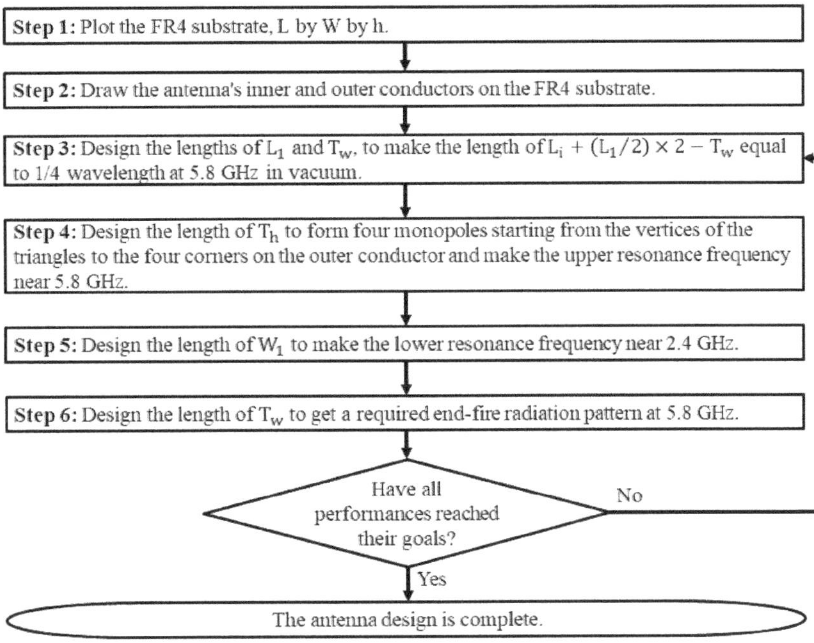

Figure 3. The main design steps of the proposed antenna.

2.2. *Antenna Design for 5.8 GHz*

Figure 4 shows the simulated surface current density at 5.8 GHz, where the areas with higher surface current densities are the main radiation areas. The outer conductor forms four monopole antennas from each triangle vertex to its adjacent corners. These four monopoles have the same wavelength, λ_H, and the upper resonance frequency, f_H, can be approximated using (1), where c is the speed of light in a vacuum (3×10^8 m/s).

$$f_H \cong \frac{c}{\lambda_H} = \frac{c}{(L_2 + L_3 + L_1/2)/0.23} = 5.63 \text{ GHz}, \quad (1)$$

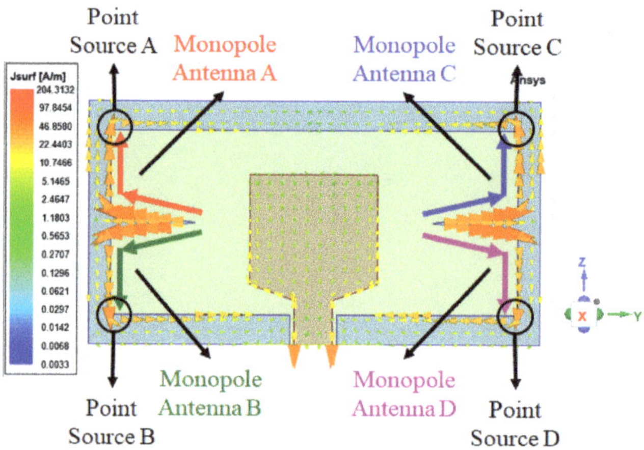

Figure 4. The simulated surface current density at 5.8 GHz.

Higher surface current densities exist at the four corners inside the outer conductor, which approximate four point sources named Element A, Element B, Element C, and Element D. The radiation pattern is mainly synthesized from these four point sources. Due to the electric field that radiates between the inner and outer conductors, the proposed antenna approximates a two-element array of two point sources, Elements 1 and 2, as shown in Figure 5.

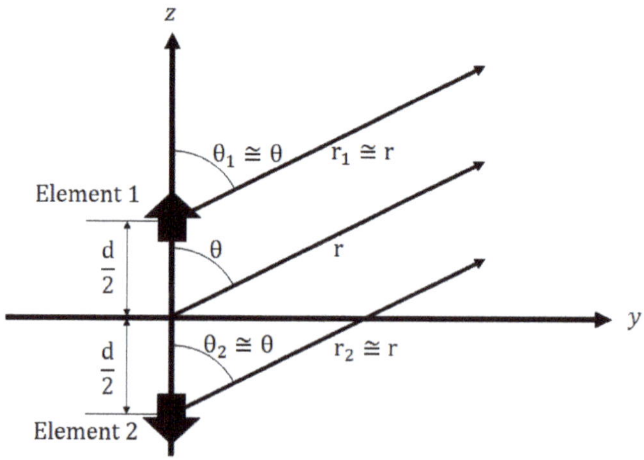

Figure 5. Far-field observation of a two-element array consisting of two point sources, Elements 1 and 2, positioned along the *z*-axis.

The radiation patterns of Elements 1 and 2 equal the sum of Elements A and C and the sum of Elements B and D, respectively. Furthermore, since the cross-sectional area of the inner conductor used as the reference plane is very small, the radiation patterns of Elements 1 and 2 approximate a vertical infinitesimal dipole. Based on the directivity of the vertical infinitesimal (Hertzian) dipole [17], the normalized pattern function of the vertical infinitesimal dipole is $\sin^2\theta$. The normalized array factor of Elements 1 and 2, $(AF_{12})_n$, is adapted from (2) [18], where k is the wave number equal to $2\pi/\lambda_{0\ 5.8G}$, and d is the distance between Elements 1 and 2, approximately equal to $L_i + (L_1/2) \times 2$. In addition, $\lambda_{0\ 5.8G}$ is the wavelength of 5.8 GHz in vacuum, θ is the elevation angle, and π is the ratio of a circle's circumference to its diameter (180°).

$$(AF_{12})_n = \cos\left[\frac{1}{2}(kd\cos\theta + \beta_d)\right] = \cos\left\{\frac{1}{2}\left[\frac{2\pi}{\lambda_{0\ 5.8G}}(L_i + (L_1/2) \times 2)\cos\theta + \beta_d\right]\right\} = \cos\left[\frac{1}{2}\left(100.92^\circ \times \cos\theta + \beta_d\right)\right], \quad (2)$$

At 5.8 GHz, the difference in phase excitation, $\beta_d = \beta_A - \beta_B = \beta_C - \beta_D$, can be derived by (3). Where β_A, β_B, β_C, and β_D are the phase excitations of Elements A, B, C, and D, respectively. In addition, β_A equals β_C, and the negative sign is the phase lag.

$$\beta_d = -\frac{2\pi}{\lambda_{0\ 5.8G}} \times [L_i + (L_1/2) \times 2 - T_w] = -90.48^\circ, \quad (3)$$

In (2), replacing β_d with -90.48°, the normalized array factor of Elements 1 and 2 can be written as (4). It has two maximums at $\theta = +26^\circ$ and $\theta = -26^\circ$, only one minimum at $\theta = +180^\circ$, and forms an end-fire array toward 0°.

$$(AF_{12})_n = \cos\left[\frac{1}{2}\left(100.92^\circ \times \cos\theta - 90.48^\circ\right)\right], \quad (4)$$

The normalized total field of the two-element array antenna, $(E_{total})_n$, is (5). It has two maximums at $\theta = +60^\circ$ and $\theta = -60^\circ$, only one minimum at $\theta = +180^\circ$, and forms an end-fire radiation toward 0°.

$$(E_{total})_n = \left\{\sin^2\theta \times (AF_{12})_n\right\}_n, \quad (5)$$

Figure 6 shows the relative power patterns of the vertical infinitesimal dipole, the array factor of Elements 1 and 2, and the total field of the two-element array antenna.

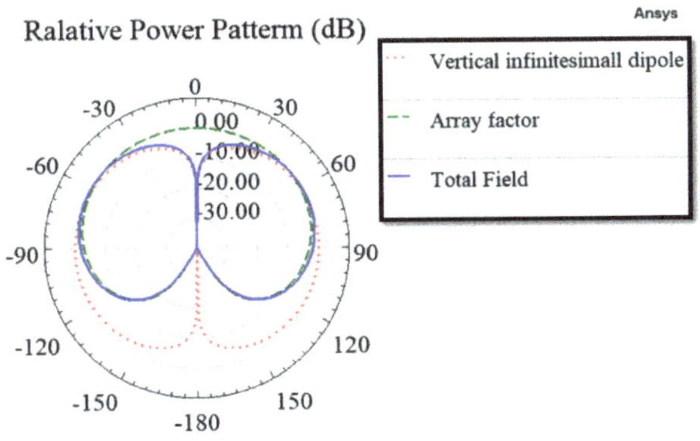

Figure 6. The relative power patterns of the vertical infinitesimal dipole, the array factor of Elements 1 and 2, and the total field of the two-element array antenna.

2.3. Antenna Design for 2.4 GHz

Figure 7 shows the simulated surface current densities at 2.4 GHz, where the areas with higher surface current densities are the main radiation areas. The two sides of the outer conductor form a pair of L-shaped monopoles, Monopoles E and F. These two monopoles have the same wavelength, λ_L, and the lower resonance frequency, f_L, can be approximated by (6).

$$f_L \cong \frac{c}{\lambda_L} = \frac{c}{\{[W_1 + (W_2/2)] + [L_i + (L_1/2) \times 2]\}/0.23} = 2.52\ \text{GHz}, \quad (6)$$

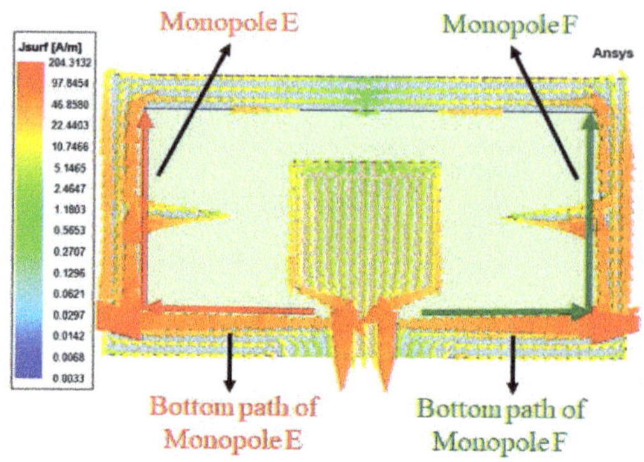

Figure 7. The simulated surface current density at 2.4 GHz.

Due to the symmetrical design of the architecture, the radiations from the two bottom paths cancel each other out. Due to the electric field that radiates between the inner and outer conductors, the proposed antenna approximates a vertical monopole antenna. In addition, since the cross-sectional area of the inner conductor reference plane is very small, the radiation pattern of the vertical monopole antenna approximates a vertical dipole with an effective length of $L_i + (L_1/2) \times 2$, equivalent to 0.122 λ_L. Its maximum radiations are near the horizontal plane.

3. Parametric Studies and Full-Wave Simulations

The proposed antenna was simulated by the ANSYS High-Frequency Structure Simulator (HFSS) [19]. From (1), f_H decreases as L_2 increases, L_2 can be derived from (7), and L_2 increases as T_h increases. Hence, f_H decreases as T_h increases.

$$L_2 = \sqrt{{T_h}^2 + \left(\frac{T_w}{2}\right)^2} = \sqrt{5.7^2 + \left(\frac{1.5}{2}\right)^2} = 5.75, \quad (7)$$

Figure 8 shows the simulated reflection coefficients for different values of T_h. The simulated data agree with (1) and (7), f_H decreases as T_h increases, and f_H is about 5.63 GHz when T_h is 5.7 mm.

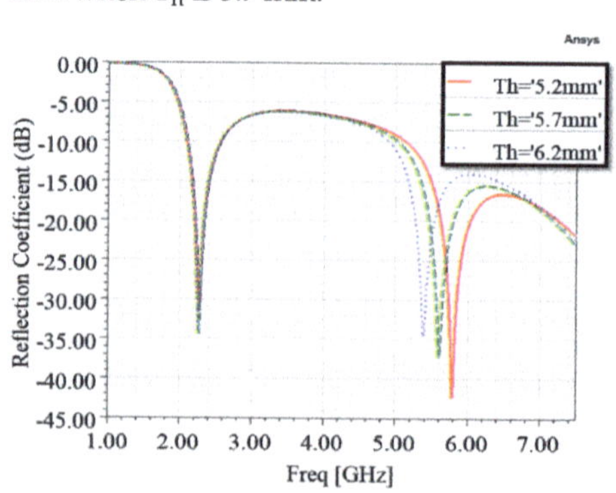

Figure 8. The simulated reflection coefficients for different values of T_h.

From (6), f_L decreases as W_1 increases, and Figure 9 shows the simulated reflection coefficients for different values of W_1. The simulated data agree with (6), f_L decreases as W_1 increases.

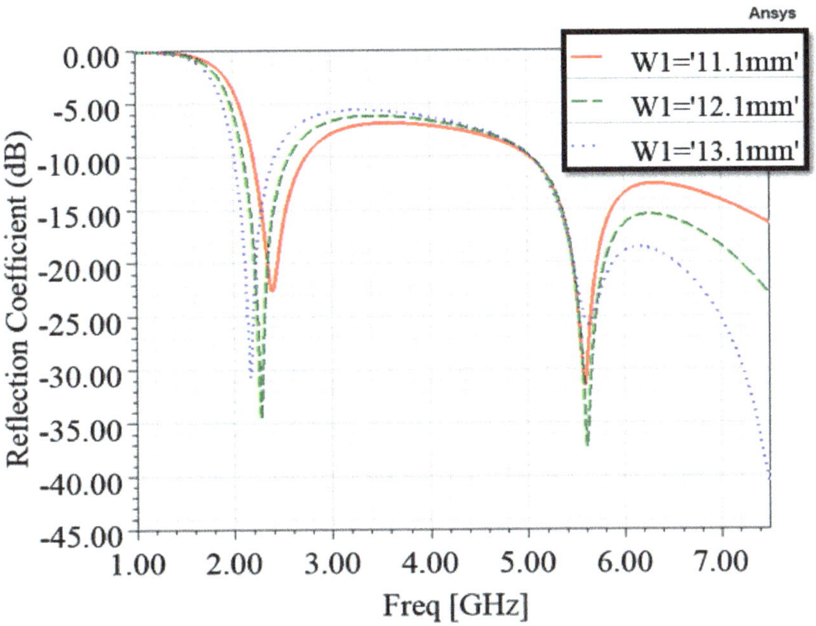

Figure 9. The simulated reflection coefficients for different values of W_1.

From (2) and (5), the normalized total field at 5.8 GHz changes as β_d changes, and from (3), β_d gets closer to -100.92° as T_w decreases closer to zero. Hence, the end-fire radiation becomes more significant as T_w decreases. Figure 10 shows the simulated vertically polarized radiation patterns at 5.8 GHz in the y–z plane for different values of T_w. The simulated data are in agreement with (2), (3), and (5), and the end-fire radiation becomes more significant as T_w decreases.

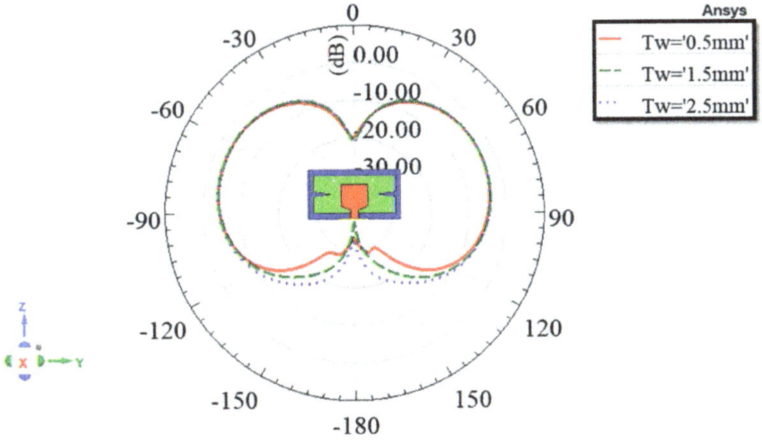

Figure 10. The simulated vertically polarized radiation patterns at 5.8 GHz in the y–z plane for different values of T_w.

4. The Comparisons of Simulated Data and Measured Results of The Prototype

Figure 11 shows the prototype of the proposed antenna with an SMA connector [20] soldered to it, where the hand-soldered tin was not incorporated into the simulation. The antenna was measured by the Atenlab A3 compact-type OTA measurement system [21] and compared with its simulated data. Figure 12a–c show the measurement setup for radiation patterns in the x–y, x–z, and y–z planes. The proposed antenna was placed on a turntable and connected to the coaxial cable of the measurement system, which was not incorporated into the simulation. Figure 12a shows that the proposed antenna was connected directly to the coaxial cable of the measurement system, and the cable was parallel to the z-axis of the proposed antenna. Figure 12b,c show that the proposed antennas were connected to the coaxial cable of the measurement system via a right-angle SMA adapter [22], and the cable was perpendicular to the z-axis of the proposed antennas.

Figure 11. The prototype of the proposed antenna with an SMA connector soldered to it.

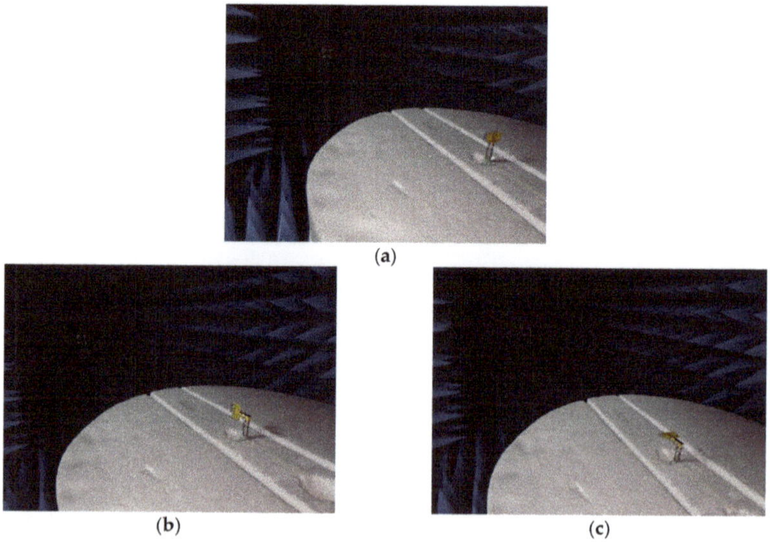

(a)

(b) (c)

Figure 12. The measurement setup for radiation patterns. (**a**) x–y plane; (**b**) x–y plane; (**c**) x–y plane.

4.1. Reflection Coefficients

Figure 13 compares the simulated and measured reflection coefficients of the proposed antenna. The simulated curve shows that the proposed antenna can meet $|S_{11}| \leq -10$ dB from 2.11 to 2.58 GHz and 5.06 to 7.5 GHz, where the fractional bandwidths are 20.04%

and 38.05%, respectively. The measured result agrees with the simulated one and can meet Wi-Fi 2.4 and 5.8 GHz requirements for small UAV applications.

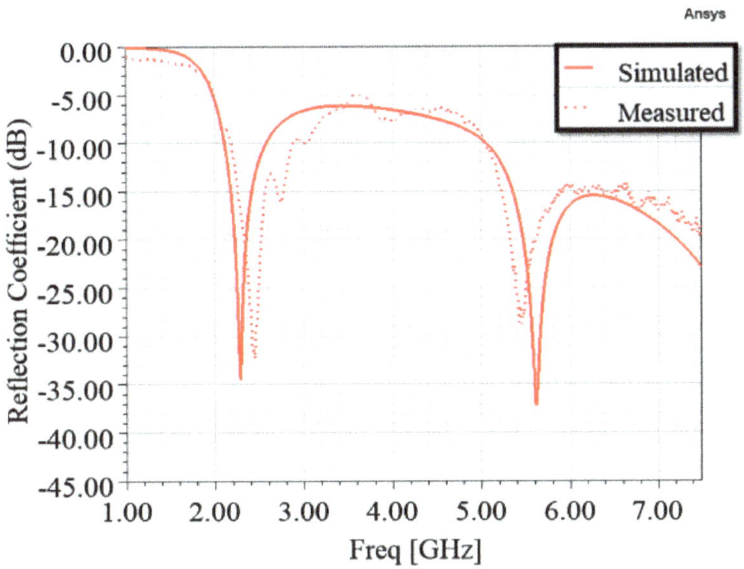

Figure 13. The comparison of the simulated and measured reflection coefficients of the proposed antenna.

4.2. Radiation Patterns

Figures 14–16 show the comparisons of the simulated and measured radiation patterns of the proposed antenna in the x–y, x–z, and y–z planes. The simulated data show that the proposed antenna has good vertically polarized radiation. Figure 14a,b show the radiation patterns of the proposed antenna in the x–y plane at 2.4 and 5.8 GHz, respectively, and the vertically polarized radiations are quasi-omnidirectional. Due to the hand-soldered tins on the main radiation areas of 2.4 GHz, the measured horizontally polarized radiation is larger than the simulated one, as shown in Figure 14a.

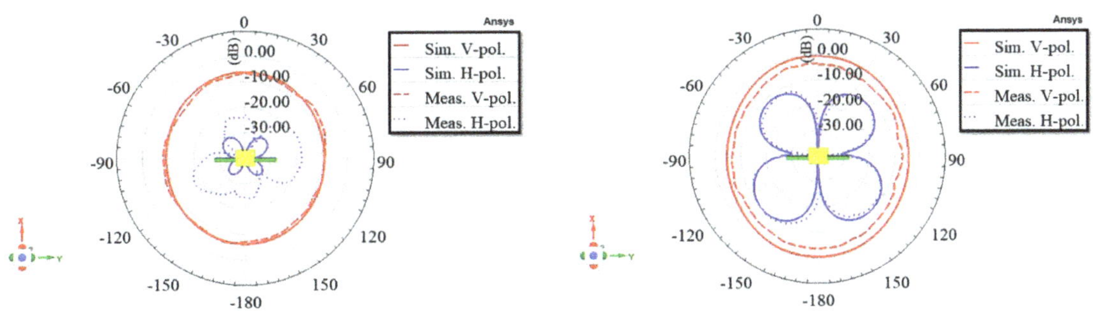

(**a**) (**b**)

Figure 14. Comparison of the proposed antenna's simulated and measured radiation patterns in the x–y plane. (**a**) 2.4 GHz; (**b**) 5.8 GHz.

In the x–z and y–z planes, due to the cable effect, the measured horizontally polarized radiations are much larger than the simulated data, as shown in Figures 15 and 16. In addition, due to the cable effect, the measured vertically polarized radiation is larger than the simulated one at −120° to +120° in the y–z plane at 5.8 GHz, as shown in Figure 16b.

The simulated vertically polarized radiation patterns at 2.4 GHz are quasi-omnidirectional, and the maximum radiations in the x–z and y–z planes are near the horizontal plane, as shown in Figures14a, 15a and 16a. The simulated vertically polarized radiation patterns at 5.8 GHz have stronger radiation in the upper hemisphere (below the UAV), as shown in Figures 15b and 16b. Its maximum radiations in the x–z and y–z planes are in the upper hemisphere and between 14 and 29° from the horizontal plane.

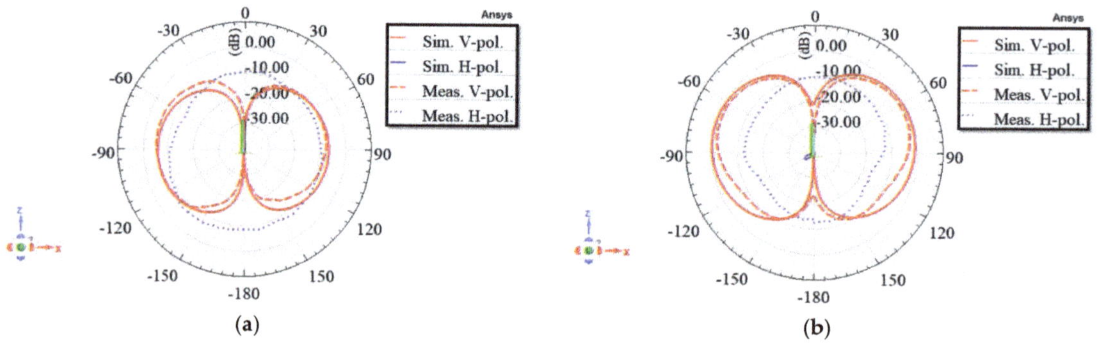

Figure 15. Comparison of the proposed antenna's simulated and measured radiation patterns in the x–z plane. (**a**) 2.4 GHz; (**b**) 5.8 GHz.

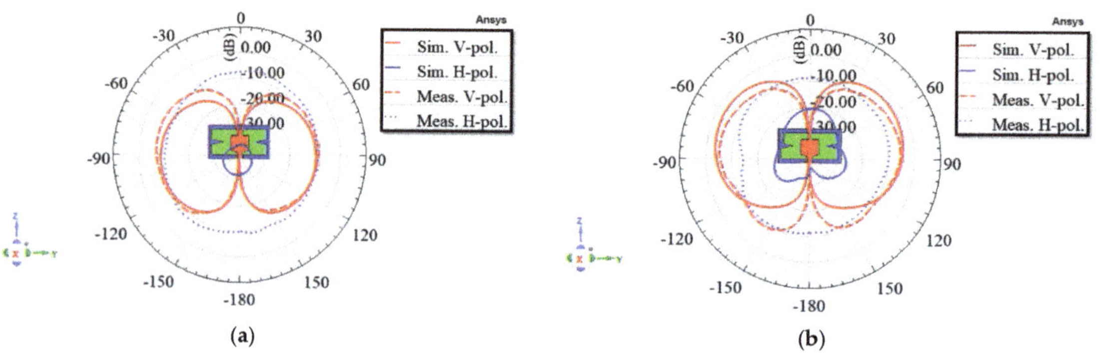

Figure 16. Comparison of the proposed antenna's simulated and measured radiation patterns in the y–z plane. (**a**) 2.4 GHz; (**b**) 5.8 GHz.

Table 2 displays the characteristics of the simulated vertically polarized radiation patterns. Ignoring the influence caused by the hand-soldered tin and the measurement system's cable effect, all measured radiation pattern results agree with simulated data. In E-planes, the beamwidths of relative −10 dB at 2.4 GHz are no less than 142°. Moreover, at 5.8 GHz, maximum gain angles are between 17 and 29° off the horizontal.

Table 3 compares the proposed antenna and published works on planar antennas. The proposed planar antenna has the following advantages. First, it has a single-sided structure, which has a relatively low cost to make small UAVs more affordable. Second, it has a minimum electrical length (L) of 0.132 λ_0 and a low profile (h) of 0.0128 λ_0 to meet the portability and aerodynamic requirements for small UAV applications. Third, at 5.8 GHz, it has stronger radiation in the upper hemisphere (below the small UAV) to reduce interference from rare-use directions. Its maximum radiations are below the UAV and between 14 and 29° from the horizontal plane to provide more antenna gain when the small UAV flies farther in urban areas. At 2.4 GHz, it has quasi-omnidirectional radiation to ensure a stable link in all directions. Its maximum radiations are near the horizontal plane to provide more antenna gain when the small UAV flies farther in open areas. Finally, it has good vertically polarized radiation at 2.4 and 5.8 GHz for long-distance applications.

Table 2. Characteristics of the simulated vertically polarized radiation patterns.

Unit: Degree			Y–z Plane				x–z Plane			
			Left Plane		Right Plane		Left Plane		Right Plane	
2.4 GHz	Maximum gain angle		−98		98		−95		91	
	Relative −3 dB	Beam angle	−142	−49	48	142	−139	−49	46	134
		Beamwidth	93		94		90		88	
	Relative −10 dB	Beam angle	−165	−20	19	165	−165	−21	18	160
		Beamwidth	145		146		144		142	
5.8 GHz	Maximum gain angle		−61		62		−76		73	
	Relative −3 dB	Beam angle	−102	−29	29	103	−118	−39	37	114
		Beamwidth	72.96		74		79		77	
	Relative −10 dB	Beam angle	−127	−11	11	129	−152	−16	14	148
		Beamwidth	116		118		136		134	

Table 3. The comparison of the proposed antenna and published works on planar antennas.

Ref.	Structure	Operating Frequency	Radiation Pattern for Urban Areas	Radiation Pattern for Open Areas	Electrical Dimensions (L×W×h) (λ_0)
[11]	Double-sided	0.86 to 1.51		Yes	0.3297 × 0.0344 × 0.0046
[12]	Double-sided	3.9 to 5		Yes	2.3400 × 0.1820 × 0.0260
[13]	Single-sided	3.5/3.7/4.5/5/5.7	Yes	Yes	0.4690 × 0.1470 × 0.0092
[14]	Single-sided	2.4 to 5.8		Yes	0.1600 × 0.2000 × 0.0128
This work	Single-sided	2.4/5.8	Yes	Yes	0.1320 × 0.2424 × 0.0128

5. Conclusions

In summary, we proposed a compact planar Wi-Fi antenna with optimized radiation patterns for small UAV applications in urban and open areas. The proposed antenna was prototyped and measured, and the measured results agree with the simulated data. It has end-fire radiation to reduce the interference from rare-use directions at 5.8 GHz and quasi-omnidirectional radiation to ensure a stable link in all directions at 2.4 GHz. The directions of the maximum radiation in each elevation plane at 5.8 and 2.4 GHz provide more antenna gain when the small UAV flies farther in urban and open areas. In future work, the proposed antenna will be mounted on a small UAV for field experiments.

Author Contributions: Conceptualization, Y.-L.Y. and D.-B.L.; methodology, Y.-L.Y. and D.-B.L.; software, Y.-L.Y.; validation, Y.-L.Y. and D.-B.L.; formal analysis, Y.-L.Y. and D.-B.L.; investigation, Y.-L.Y. and D.-B.L.; resources, D.-B.L.; data curation, Y.-L.Y. and D.-B.L.; writing—original draft, Y.-L.Y.; writing—review and editing, D.-B.L.; visualization, Y.-L.Y.; supervision, D.-B.L.; project administration, D.-B.L. All authors have read and agreed to the published version of the manuscript.

Funding: This work was supported by the Ministry of Science and Technology of Taiwan under Grant MOST 110-2221-E-011-052.

Institutional Review Board Statement: Not applicable.

Informed Consent Statement: Not applicable.

Data Availability Statement: All the design parameters and related data are provided in this article.

Conflicts of Interest: The authors declare no conflict of interest.

References

1. Xu, C.; Liao, X.; Tan, J.; Ye, H.; Lu, H. Recent Research Progress of Unmanned Aerial Vehicle Regulation Policies and Technologies in Urban Low Altitude. *IEEE Access* **2020**, *8*, 74175–74194. [CrossRef]
2. Balanis, C.A. *Antenna Theory: Analysis and Design*, 4th ed.; Wiley: New York, NY, USA, 2016; Chapter 2; p. 89.
3. Barani, N.; Harvey, J.F.; Sarabandi, K. Fragmented Antenna Realization Using Coupled Small Radiating Elements. *IEEE Trans. Antennas Propag.* **2018**, *66*, 1725–1735. [CrossRef]
4. Pfeiffer, C.; Dagefu, F.T. A Compact Electrically Tunable VHF Antenna. In Proceedings of the 2019 IEEE International Symposium on Antennas and Propagation and USNC-URSI Radio Science Meeting, Atlanta, GA, USA, 7–12 July 2019; pp. 1549–1550. [CrossRef]
5. Sun, L.; Sun, B.; Yuan, J.; Tang, W.; Wu, H. Low-Profile, Quasi-Omnidirectional Substrate Integrated Waveguide (SIW) Multihorn Antenna. *IEEE Antennas Wirel. Propag. Lett.* **2016**, *15*, 818–821. [CrossRef]
6. Pradhan, S.; Gupta, B. High-Gain Dual-Mode Cylindrical Rectangular Patch Antenna for Airborne Applications. *IEEE Trans. Aerosp. Electron. Syst.* **2022**, *58*, 4168–4179. [CrossRef]
7. Duran-Pardo, A.V.; Navarro-Mendez, D.V.; Carrera-Suarez, L.F.; Baquero-Escudero, M. Conformal Patch Antennas for an UAV employed in Agricultural Applications. In Proceedings of the 2020 IEEE International Symposium on Antennas and Propagation and North American Radio Science Meeting, Montreal, QC, Canada, 5–10 July 2020; pp. 285–286. [CrossRef]
8. Han, Y.; Hu, K.; Zhao, R.; Gao, Y.; Dai, L.; Fu, Y.; Zhou, B.; Yuan, S. Design of Combined Printed Helical Spiral Antenna and Helical Inverted-F Antenna for Unmanned Aerial Vehicle Application. *IEEE Access* **2020**, *8*, 54115–54124. [CrossRef]
9. Wang, C.; Yuan, B.; Shi, W.; Mao, J. Low-Profile Broadband Plasma Antenna for Naval Communications in VHF and UHF Bands. *IEEE Trans. Antennas Propag.* **2020**, *68*, 4271–4282. [CrossRef]
10. Nosrati, M.; Jafargholi, A.; Pazoki, R.; Tavassolian, N. Broadband Slotted Blade Dipole Antenna for Airborne UAV Applications. *IEEE Trans. Antennas Propag.* **2018**, *66*, 3857–3864. [CrossRef]
11. Chou, H.-T.; Lin, D.-B.; Chiu, H.-L. Planar Dual-Mode Dipole Antenna Formed by Artificial Microstrip Arms Loaded with Multiple H-Slots for Broadband Operation of Vertically Polarized Radiation on UAV Platform. *IEEE Trans. Antennas Propag.* **2022**, *70*, 7869–7877. [CrossRef]
12. Yan, Y.-D.; Jiao, Y.-C. Omnidirection Vertically Polarized Antenna on Unmanned Aerial Vehicle. In Proceedings of the 2018 12th International Symposium on Antennas, Propagation and EM Theory (ISAPE), Hangzhou, China, 3–6 December 2018; pp. 1–3. [CrossRef]
13. Sumi, M.; Suzuki, Y. A Wideband Single-Sided Folded-Off-Center-Fed Dipole Antenna for 4G/5G/Wi-Fi M2M/IoT Applications and UAVs. In Proceedings of the 2021 IEEE Conference on Antenna Measurements & Applications (CAMA), Antibes Juan-les-Pins, France, 15–17 November 2021; pp. 3–4. [CrossRef]
14. Lin, D.-B.; Tang, I.-T.; Wei, Y.-J.; Lin, H.-P. A compact slot antenna with CPW-fed for IEEE 802.16-2004 applications. In Proceedings of the 2007 IEEE Antennas and Propagation Society International Symposium, Honolulu, HI, USA, 9–15 June 2007; pp. 4753–4756. [CrossRef]
15. Reddy Maddikunta, P.K.; Hakak, S.; Alazab, M.; Bhattacharya, S.; Gadekallu, T.R.; Khan, W.Z.; Pham, Q.-V. Unmanned Aerial Vehicles in Smart Agriculture: Applications, Requirements, and Challenges. *IEEE Sens. J.* **2021**, *21*, 17608–17619. [CrossRef]
16. *IEEE Std 145-2013 (Revision of IEEE Std 145-1993)*; IEEE Standard for Definitions of Terms for Antennas. IEEE: New York, NY, USA, 2014; pp. 1–50. [CrossRef]
17. David, K.C. *Fundamentals of Engineering Electromagnetics*; Addison-Wesley: Menlo Park, CA, USA, 1993; Chapter 10; pp. 432–433.
18. Balanis, C.A. *Antenna Theory: Analysis and Design*, 4th ed.; Wiley: New York, NY, USA, 2016; Chapter 6; p. 287.
19. ANSYS Incorporation, High-Frequency Structure Simulator (HFSS), Version 2022 R1, 7 November 2022. Available online: https://www.ansys.com/products/electronics/ansys-hfss (accessed on 16 June 2023).
20. SMA PCB End Launch Straight Jack 50 Ohm, Amphenol Corporation. January 2023. Available online: https://www.amphenolrf.com/901-10309.html (accessed on 16 June 2023).
21. Atenlab A3 Measurement System, PCB GraphTech Pte. November 2022. Available online: https://pcbgt.com.sg/atenlab-a3 (accessed on 16 June 2023).
22. SMA Jack to SMA Plug Adapter 50 Ohm Right Angle, Amphenol Corporation. January 2023. Available online: https://www.amphenolrf.com/132172.html (accessed on 16 June 2023).

applied sciences

MDPI

Article

Wideband, High-Gain, and Compact Four-Port MIMO Antenna for Future 5G Devices Operating over Ka-Band Spectrum

Sayed Aqib Hussain [1], Fatma Taher [2], Mohammed S. Alzaidi [3], Irshad Hussain [4], Rania M. Ghoniem [5,*], Mohamed Fathy Abo Sree [6] and Ali Lalbakhsh [7,8,*]

1 Department of Electrical Engineering, Bahria University, Islamabad 44000, Pakistan
2 College of Technological Innovation, Zayed University, Dubai 19282, United Arab Emirates
3 Department of Electrical Engineering, College of Engineering, Taif University, P.O. Box 11099, Taif 21944, Saudi Arabia
4 Department of CS and IT, University of Sargodha, Sargodha 40100, Pakistan
5 Department of Information Technology, College of Computer and Information Sciences, Princess Nourah bint Abdulrahman University, P.O. Box 84428, Riyadh 11671, Saudi Arabia
6 Department of Electronics and Communications Engineering, Arab Academy for Science, Technology and Maritime Transport, Cairo 11799, Egypt
7 School of Engineering, Macquarie University, Sydney, NSW 2109, Australia
8 School of Electrical and Data Engineering, University of Technology Sydney, Ultimo, NSW 2007, Australia
* Correspondence: rmghoniem@pnu.edu.sa (R.M.G.); ali.lalbakhsh@mq.edu.au (A.L.)

Abstract: In this article, the compact, ultra-wideband and high-gain MIMO antenna is presented for future 5G devices operating over 28 GHz and 38 GHz. The presented antenna is designed over substrate material Roger RT/Duroid 6002 with a thickness of 1.52 mm. The suggested design has dimensions of 15 mm × 10 mm and consists of stubs with loaded rectangular patch. The various stubs are loaded to antenna to improve impedance bandwidth and obtain ultra-wideband. The resultant antenna operates over a broadband of 26.5–43.7 GHz, with a peak value of gain >8 dBi. A four-port MIMO configuration is achieved to present the proposed antenna for future high data rate devices. The MIMO antenna offers isolation <−30 dB with ECC of <0.0001. The antenna offers good results in terms of gain, radiation efficiency, envelop correlation coefficient (ECC), mean effective gain (MEG), diversity gain (DG), channel capacity loss (CCL), and isolation. The antenna hardware prototype is fabricated to validate the performance of the suggested design of the antenna achieved from software tools, and good correlation between measured and simulated results is observed. Moreover, the proposed work performance is also differentiated with literature work, which verifies that the suggested work is a potential applicant for future 5G compact devices operating over wideband and high gain.

Keywords: wideband antenna; compact size; high gain; Ka-band; 5G

Citation: Hussain, S.A.; Taher, F.; Alzaidi, M.S.; Hussain, I.; Ghoniem, R.M.; Sree, M.F.A.; Lalbakhsh, A. Wideband, High-Gain, and Compact Four-Port MIMO Antenna for Future 5G Devices Operating over Ka-Band Spectrum. *Appl. Sci.* **2023**, *13*, 4380. https://doi.org/10.3390/app13074380

Academic Editor: Naser Ojaroudi Parchin

Received: 3 March 2023
Revised: 24 March 2023
Accepted: 27 March 2023
Published: 30 March 2023

1. Introduction

Due to high data rate and high link capacity, the multiple input and multiple output (MIMO) antenna is widely used in modern wireless communication systems, especially operating over millimeter wave spectrum for 5G devices [1,2]. It also has advantages to avoid seamless connectivity, as in the MIMO system, multiple antennas are installed at transmitter and receiver end. Due to these benefits, a MIMO antenna is a hot trend topic in current research on designing antenna for 5G and 6G communication devices [3,4]. The requirement of the MIMO antenna altered the requirement of antenna designing, as moving towards the MIMO system, the simplified geometry along with low-profile and compact size will be most beneficial [5,6].

On other hand, the performance parameters are also important to examine and analyze, as high data rate and low latency is demanded to facilitate a large number of users with good services. These demands need an antenna to operate over wideband and offer high gain along with radiation efficiency [7,8]. In the case of the MIMO antenna system, the MIMO parameters are also important to be considered. The mutual coupling, envelop correlation coefficient (ECC), and diversity gain (DG) and many others are key parameters, which will be analyzed [9,10].

In literature, the researcher has present a number of monopole antennas operating over 28 GHz [11–20], and various MIMO systems operational over 28 GHz for wireless communication systems [21–32]. The author of the selected literature reported in [11–14] presents the antenna with compact size and simplified geometry. The setback of these designs is that the report work has a lack of measured results or low gain or radiation efficiency. The antenna reported in [15–19], offers high gain and has simplified geometry. The demerit of these works is narrow operational bandwidth or large size. The requirement of 5G devices is the antenna having a compact size, low profile, and simplified geometry along with wideband and high gain.

According to the requirements of the 5G communication system, the MIMO antenna, which offers a high number of gains, is reported in literature. In [19], the MIMO antenna with improved isolation is proposed by inserting electromagnetic band gap (EBG) between MIMO elements. The reported four-port MIMO antenna operates at 28/38 GHz with a bandwidth of 4 GHz and 2 GHz. Although the antenna is operational over high gain and broadband, it has complex geometry due to inserting EBG cells. A two-port MIMO antenna with simple geometry is reported in [21]. The geometry consists of defected ground structure (DGS) to improve the performance. This work has the demerit of a narrow bandwidth of 1.1 GHz. Another work that offers high gain and wideband is reported in [22]. The antenna consists of eight ports, which lead the reported work with large dimensions.

Compact and wideband antenna operating on mm wave application is reported in [23]. The antenna has a compact size of 24 mm × 20 mm × 1.85 mm and operates over a wideband of 33–44 GHz, but a setback of low value of ECC (that is 0.1). A wideband and high-gain antenna with a bandwidth of 22–50 and peak gain of 15 dBi is reported in [24]. The antenna has complex geometry, as array is adopted to achieve the high gain. In [25], an antenna is presented for the 5G communication system, which offers a bandwidth of 27.1–28.1. The antenna also has large dimensions of 110 mm × 55 mm × 1.6 mm. A compact antenna with an overall size of 12.5 mm × 12.5 mm × 0.8 mm and peak gain of 6 dBi is reported in [26]. The antenna consists of small size and is operational over high gain, but offers low value of other MIMO performance parameters.

A two-port MIMO antenna with geometrical dimensions of 60 mm × 100 mm × 0.965 mm and operating over a bandwidth of 27.6–28.3 GHz and ECC of 0.134. This work has the advantage of simplified geometry, but the demerit of large dimensions, narrow bandwidth, and low value of ECC [27]. A compact MIMO antenna with dimensions of 18.5 mm × 18.5 mm is reported in [28]. The antenna is compact in size and offers a wideband of 4 GHz, but complex geometry due to the multi-layer structure. Another compact antenna with an overall size of 48 mm × 31 mm × 0.254 mm is reported in [29]. The antenna has the advantage of compact size and offers a wideband of 26–31 GHz, but the demerit of high isolation of −21 dB; [30] reports a broad and strong gain. The antenna provides a high gain of 12 dBi and a wideband of 23–40 GHz. This design's setback is substantial, measuring 80 mm by 80 mm by 1.57 mm. The dimensions of a small, high-gain, broadband antenna that operates throughout the frequency range of 25.5 to 30 GHz and has a peak gain of 8.75 dBi are given in [31]. The antenna is small, wideband, and high-gain; however, the authors failed to mention the importance of ECC, and its shape is complicated.

It can be observed from the above discussion and the rest of the literature that the MIMO antenna configuration of the patch antenna is adopted due to its numerous advantages and applications. The ratch antenna with MIMO configuration has an application in smart mobile phones [32] with a high performance parameter and simple geometry, and also in wearable applications when the radiator is placed over flexible substrate material [33]. To obtain the good results and requirements for 5G applications, various techniques are adopted to improve the performance of the MIMO antenna [34,35]. The most important parameter is mutual coupling between MIMO elements, and some design meta-material is inspired to improve the mutual coupling [36]. The decoupling structure, also called parasitic patch, is loaded between antenna elements to improve isolation [37]. The corner-fed technique [38] and multi-feeding method [39] are also used to get the desired results operational over the 5G spectrum. Moreover, a low-cost MIMO antenna configuration operating over wideband is also adopted for IoT applications [40]. In addition to the patch antenna, Yagi antenna with MIMO configuration is also adopted in the literature for 5G applications [41].

From the above literature review, it is clear that there is still a research gap to design an antenna with compact size, a low profile, and simplified geometry. The antenna should operate over a wideband and offer high gain and radiation efficiency. In this article, the antenna is presented to overcome the demands of a 5G wireless communication system. The antenna offers ultra-wideband and high peak gain. The MIMO parameters also lie according to requirements of any MIMO system. In the rest of the paper, the single element design of the antenna is studied along with results in the form of a S-parameter, gain verses frequency plot, radiation pattern, and radiation efficiency. Afterwards, the MIMO configuration is analyzed, and comparison is performed between measured and simulated results. Last, the comparison table is added to compare the results of proposed antenna with literature along with conclusion and references.

The novelty of proposed work is:

- The compact size and simplified geometry;
- Wide operational band and high gain;
- Low mutual coupling between MIMO elements;
- Good values of MIMO parameters, such as, ECC, CCL, DG, and MEG.

2. Wideband Antenna Designing

2.1. Design Methodology

Figure 1 depicts the structure of the suggested design of an antenna operating over 28 GHz. The antenna is designed over substrate material Roger RT/Duroid 6002 with a loss tangent of 0.0012 and relative permittivity of 2.2. The antenna has an overall size of L × W × H = 15 mm × 10 mm × 1.52 mm. The antenna has a simplified structure with a rectangular and circular patch loaded with arm-shaped stubs. These stubs are loaded to the antenna to improve antenna performance in terms of bandwidth and return loss. Moreover, the antenna is designed on the commercially available electromagnetic (EM) software tool high frequency structure simulator (HFSSv9). The optimized value of antenna parameters is given below. W = 10; L = 15; H = 1.52; A = 2; B = 2.58; C = 4; D = 4; E = 1; F_X = 6; F_Y = 0.75; and R = 2. All units are in millimeters (mm).

To obtain the optimized results, few design steps are performed. Initially, an antenna with rectangular stub is design in first step for 28/38 GHz wireless applications. The length and width of the antenna are obtained from the below formula [42,43]:

$$L_{eff} = \frac{c}{2F\sqrt{\varepsilon reff}} \tag{1}$$

$$\varepsilon_{reff} = \frac{\varepsilon r + 1}{2} + \frac{\varepsilon r - 1}{2}\left[\frac{1}{1 + 12\frac{H}{W}}\right] \tag{2}$$

$$\Delta L = 0.412H\frac{(\varepsilon reff + 0.3)\left(\frac{W}{H} + 0.264\right)}{(\varepsilon reff - 0.258)\left(\frac{W}{H} + 0.8\right)} \tag{3}$$

The resultant antenna operates at 28 GHz and 37 GHz with a return loss of −10.25 dB and −8.5 dB, respectively. After that, in the second step, a circular radiator of radius

R = 2 mm is loaded to the antenna in order to improve the return loss, which will help in the enhancement of bandwidth. The antenna obtained from this step offers multiple resonances at 25 GHz, 29 GHz, 38 GHz, and 42 GHz with return losses of −12.5 dB, −15 dB, −16 dB, and −14.35 dB, respectively. In the third stage, an arm-shaped stub is added as given in the figure. This step improves the return loss of the antenna, which leads the antenna to operate over a wideband. In the final stage, the length of the arm is improved to get further improvement in the bandwidth of the antenna. The improvement is possible until the length of 4 mm. The antenna obtained from the final stage offers ultra-wideband from 26.5 to 43.7 GHz, as given in Figure 2.

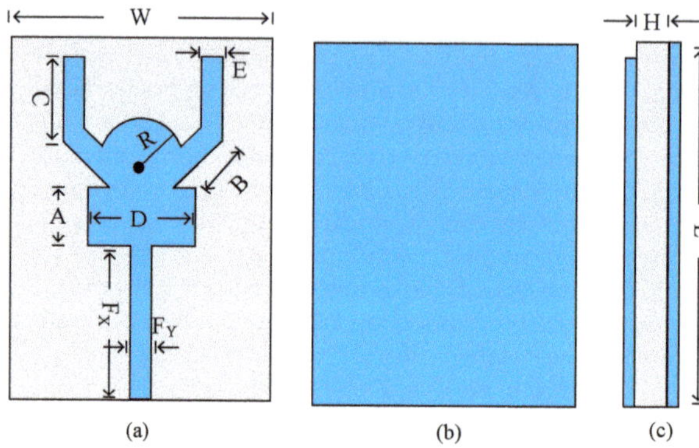

Figure 1. Geometrical configuration of proposed a dual-band millimeter wave antenna: (**a**) top view, (**b**) back view, and (**c**) side view.

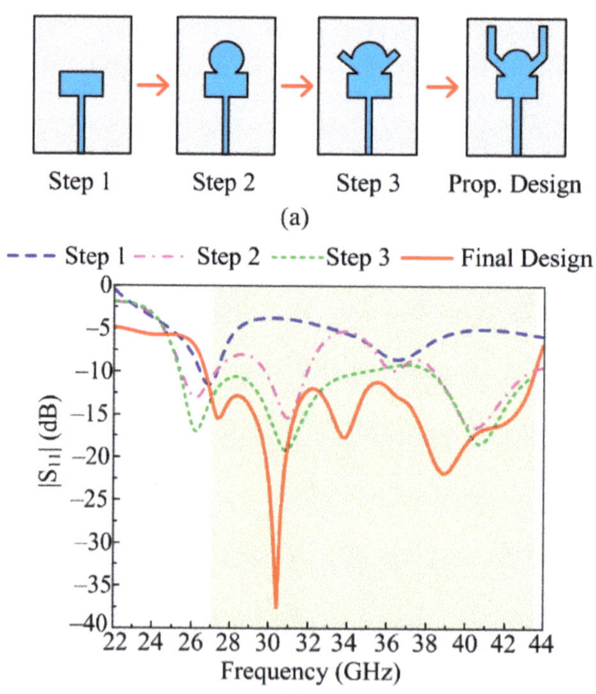

Figure 2. (**a**) Designing steps; (**b**) corresponding results.

Parametric analysis of key parameters is performed to get an optimized value. For the proposed antenna, the parametric analysis radius of the circular patch and length of the lower stub is analyzed. At an optimal value of R = 1.8 mm, the antenna offers the ultra-wideband of 26.5–43.7 GHz with a minimum value of return loss around −38 dB. If the value is increased to 2.0 mm, the antenna stops operating over wideband and only offers dual band with a minimum return loss of −18 dB and −11 dB at a resonant frequency of 29.5 GHz and 38.5 GHz, respectively. If the value is decreased to 1.6 mm, again the return loss and bandwidth is compromised, as given in Figure 3a. Another key parameter is the length of lower stub D. At optimal value of D = 4 mm, the antenna offers the required operational band. If the value is fixed at D = 6 mm, the antenna offers a 30–37 GHz bandwidth with a return loss of around −15 dB. If the value is fixed at D = 2 mm, the antenna offers 31.5–42.5 GHz with a return loss of around −20 dB, as given in Figure 3b.

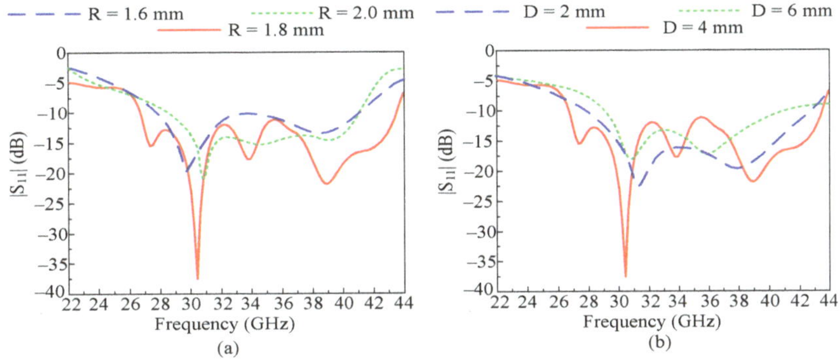

Figure 3. Parametric analysis of (**a**) the radius of the circular patch and (**b**) length of the lower stub.

2.2. Results of Unit Elements

To validate the performance of the antenna, the important parameters are analyzed. For further clarification, the hardware prototype is fabricated to compare with simulated results, as shown in Figure 4. In this section, various performance parameters are discussed and compared with measured results.

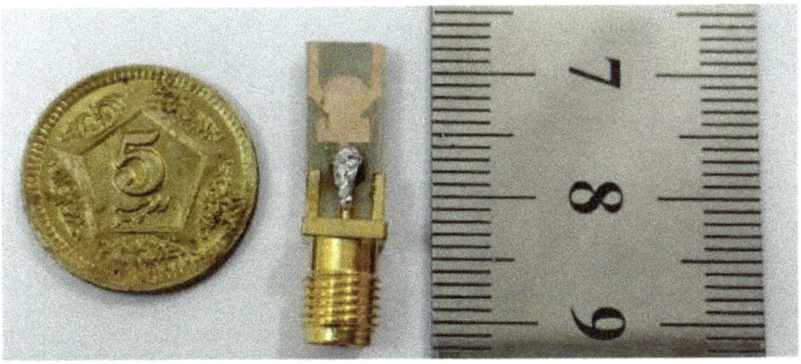

Figure 4. Fabricated prototype of wideband antenna.

2.2.1. $|S_{11}|$

Figure 5 represents the comparison between the measured and simulated S-parameter of the proposed ultra-wideband antenna. It can be seen from the figures that the antenna offers broadband ranging from 26.5 to 43.7 GHz with resonance frequencies of 30 GHz, 34, and 38 GHz. The proposed work covers the 5G band for millimeter wave application. The antenna offers a low value of return loss with a minimum −38 dB and maximum −15 dB.

The results also show similarity between simulated and measured results, which makes the proposed work the best applicant for future 5G devices operating at broadband.

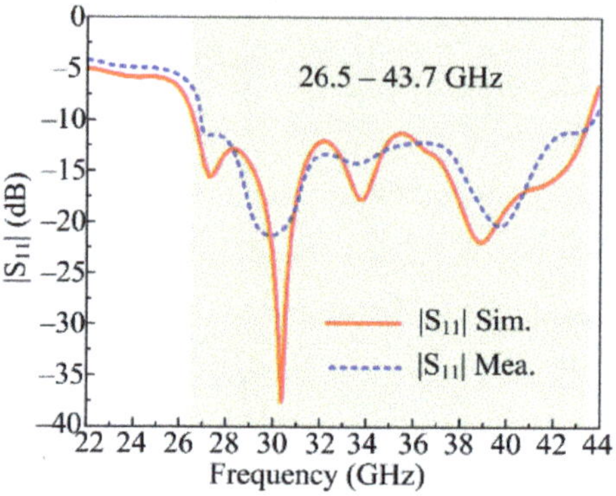

Figure 5. Comparison among simulated and experimental $|S_{11}|$ results.

2.2.2. Gain

Figure 6 shows a comparison of the proposed antenna's prototype gain and software-predicted gain. At the resonance frequency of 30 GHz, the antenna delivers a peak gain of >8 dB. It may be noted from the figures that the antenna delivers gain >6.5 dBi at an operational bandwidth of 26.5–43.7 GHz. The suggested antenna is the best candidate for future 5G devices using wideband and high gain because of the great agreement between predicted and tested outcomes.

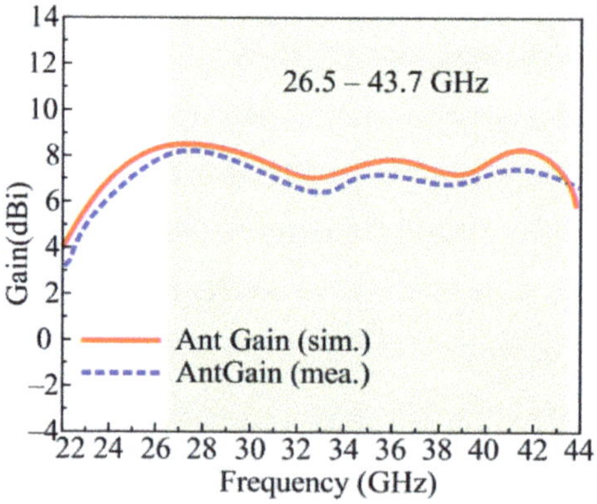

Figure 6. Comparison among simulated and experimental gain results.

2.2.3. Radiation Pattern

The proposed ultra-wideband antenna's observed and modelled radiation patterns at resonance frequencies of 30 GHz and 28 GHz are shown in Figure 7a,b. It is clear that the antenna provides both frequencies with a broad side radiation pattern in the E-plane and a slightly inclined radiation pattern in the H-plane. The tiny distortion in the radiation pattern is due to loading several stubs. Both the generated and observed radiation patterns exhibit strong correlations with one another. When testing, a little discrepancy that results from manufacturing flaws or connection loss is noticed. Because of the results and correlation between measured and simulated outcomes, future 5G devices running at broadband and high gain may be interested in the proposed study. The antenna offers a low cross polarization of <−14 dB for both resonating frequencies.

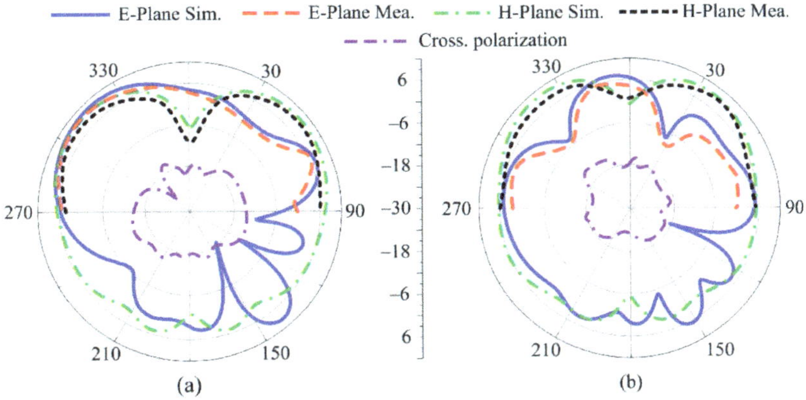

(a) (b)

Figure 7. Radiation pattern and cross-polarization of a single element at (**a**) 28 GHz and (**b**) 38 GHz.

2.2.4. Surface Current Density

Figure 8 shows the surface current distribution of the proposed antenna operating over the ultra-wideband of 26.5–43.7 GHz. It can be seen from the figures that the current is highly distributed at feedlines and arm stubs for 28 GHz and the feedline and the lower part of the antenna for 38 GHz. This phenomenon refers to that larger the effective electrical length, which proves the generation of resonances. Moreover, the large number of current distributions at arms also verifies the wideband operation of the antenna by loading these stubs.

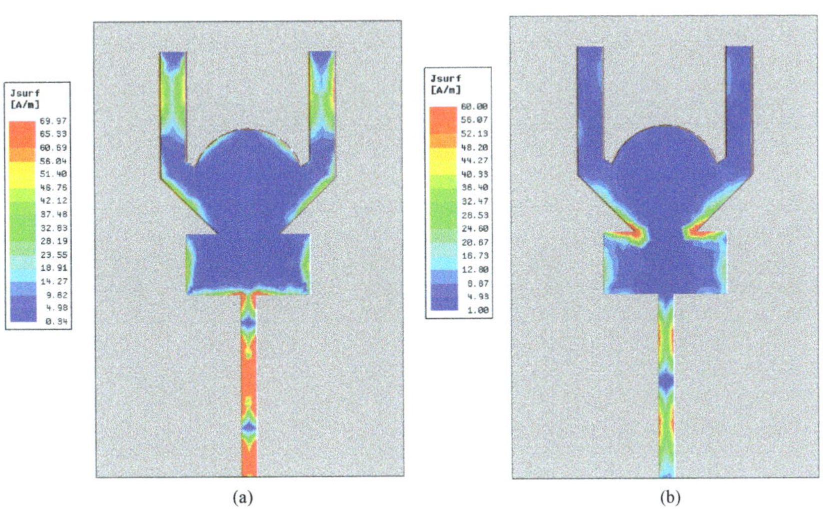

(a) (b)

Figure 8. Surface current distribution of the proposed single element of antenna (**a**) 28 GHz and (**b**) 38 GHz.

2.2.5. Radiation Efficiency

Figure 9 shows the proposed antenna's predicted radiation efficiency. The antenna has a strong radiation pattern with a 26.5–42.7 GHz working bandwidth. At resonance frequencies of 30 GHz and 38 GHz, the antenna yields peak values of 86% and 90%, respectively, with peak values of around >82% at operating bandwidth.

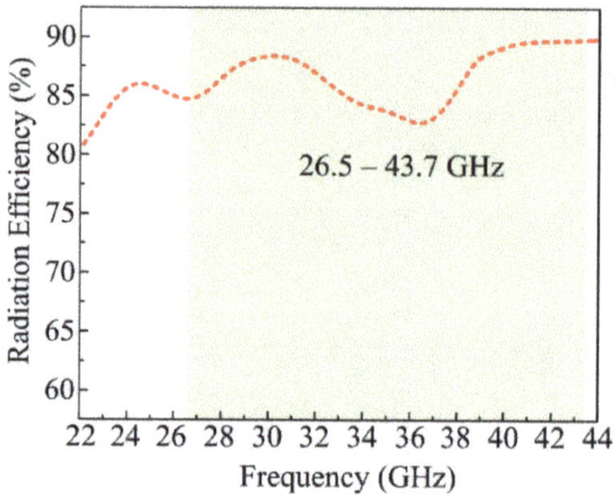

Figure 9. Simulated radiation efficiency of proposed antenna.

3. Four-Port MIMO Antenna

In this section, the MIMO configuration of the proposed antenna along with the hardware prototype is discussed. The antenna results, along with MIMO performance parameter, are also studied.

3.1. MIMO Antenna Design

Figure 10a shows the four-port MIMO antenna for ultra-wideband and high gain applications. The proposed MIMO antenna has an overall size of $M_Y \times M_X \times H$ = 27 mm × 27 mm × 1.52 mm and is designed over same material as the single element. Each unit element of the MIMO antenna has the same dimension as the single element given in Figure 1. The gap between two adjacent elements M1 = 4 mm and between two opposite elements is M_2 = 10.2 mm. Moreover, the hardware prototype is fabricated to verify the simulated results, as depicted in Figure 9b.

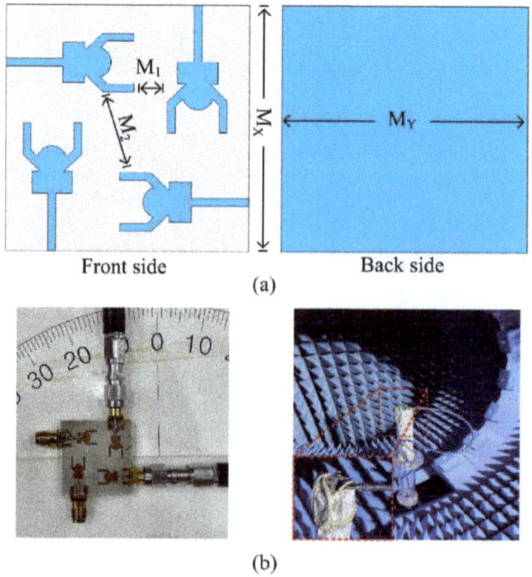

Figure 10. (**a**) Configuration of proposed MIMO antenna; (**b**) measurement setup for S-parameters and far-field parameters.

3.2. Results and Discussion

3.2.1. S-parameters

Figure 11 depicts the simulated and measured reflection and transmission coefficient of the proposed four-port ultra-wideband MIMO antenna. It can be seen that the antenna offers a broadband of 26.5–41.7 GHz with resonance frequencies of 30 GHz, 33 GHz, and 38 GHz. It can also be observed from the figures that each element of the MIMO antenna shows good correlation with each other, as well as with measured results. On other hand, the figure also shows the transmission the coefficient of the proposed antenna. According to the figures, the antenna offers minimum isolation around −55 dB and maximum isolation around −35 dB for nearby element and minimum isolation of −52 dB and maximum isolation of −32 dB with adjacent element. The results in the form of transmission and reflection coefficient and strong agreement between simulated and measured results, making the proposed MIMO antenna a good applicant for future 5G devices operating over broadband applications.

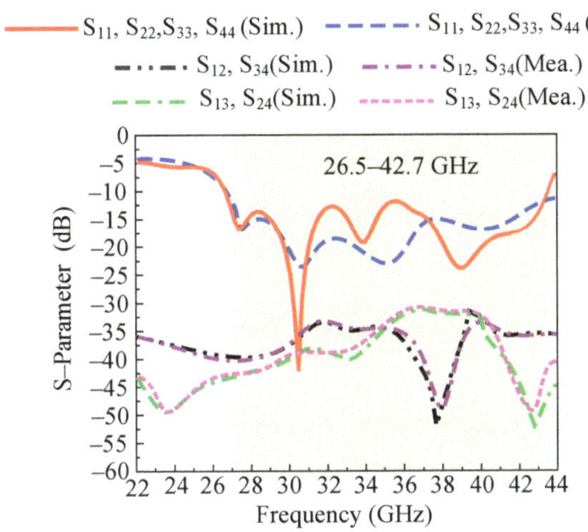

Figure 11. Comparison among S-parameters of MIMO antenna.

The minor difference between measured and simulated results may be due to:

- The fabrication tolerance of apparatus used for fabrication of antenna;
- Measurement setup tolerance due to usage of old wires;
- Connectors used in measurements as mismatching occur due to connectors at higher frequency due to increase in losses on connectors.

3.2.2. Radiation Pattern of MIMO Antenna

Figure 12 shows the ultra-wideband MIMO antenna's observed and predicted radiation pattern at resonance frequencies of 30 GHz and 38 GHz. It is clear that the antenna provides both frequencies with a broad side radiation pattern in the E-plane and a slightly inclined radiation pattern in the H-plane. The loading of several stubs is what caused the little distortion in the radiation pattern. Both the generated and observed radiation patterns exhibit strong correlations with one another. When testing, a little discrepancy that results from manufacturing flaws or connection loss is found. Because of the results and correlation between measured and simulated outcomes, future 5G devices running at broadband and high gain may be interested in the proposed study. Moreover, low cross-polarization of <−10 dB is observed for both frequencies.

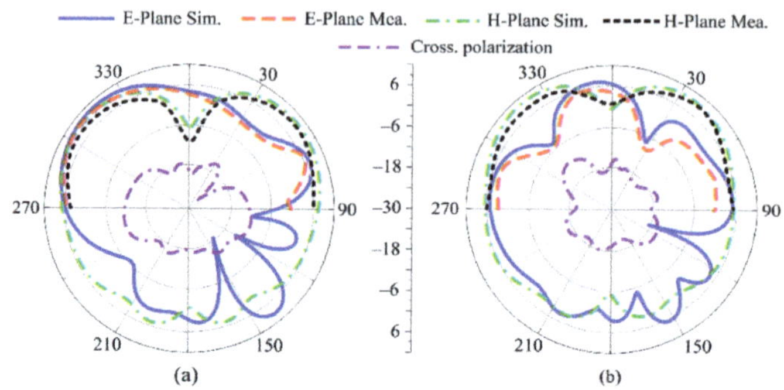

Figure 12. Radiation pattern and cross polarization of proposed MIMO antenna at (**a**) 28 GHz and (**b**) 38 GHz.

3.2.3. Electric Field Distribution

Figure 13 shows the E-field distribution of proposed MIMO antenna operating over an ultra-wideband of 26.5–42.7 GHz. It can be observed that most of the electric field is distributed among the feedline and the bottom rectangular patch for non-operating MIMO elements.

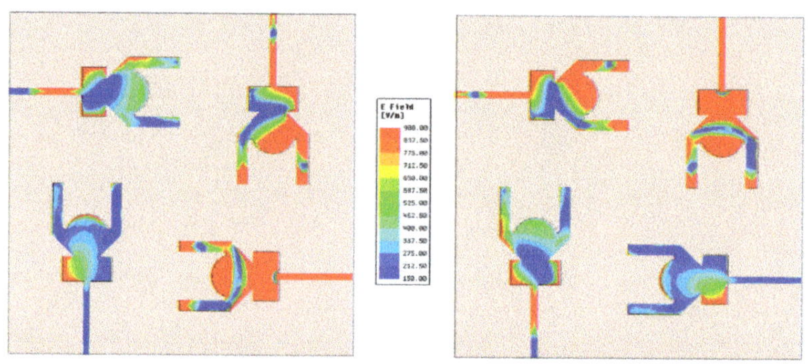

Figure 13. Electric filed distribution of proposed single element of antenna (**a**) 28 GHz and (**b**) 38 GHz.

3.2.4. Envelop Correlation Coefficient

By examining the envelope correlation coefficient (ECC), it is possible to evaluate the performance of a single MIMO antenna element. It may be measured using the S-parameter and the pattern of distant radiation. The ECCs of the adjustment element and diagonal element for the proposed ultra-wideband MIMO antenna are shown in Figure 14. As can be observed, the antenna operates over a wideband of 26.5–42.7 GHz with an ECC of 0.001. The following mathematical formulas (4) may be used to compute the ECC of a MIMO antenna [44]:

$$|\rho_e(i,j,N)| = \frac{\left|\sum_{n=1}^{N} S_{i,n}S_{n,j}\right|}{\sqrt{\left|\prod_{k(=i,j)}\left[1-\sum_{n-1}^{N} S_{i,n}S_{n,k}\right]\right.}} \quad (4)$$

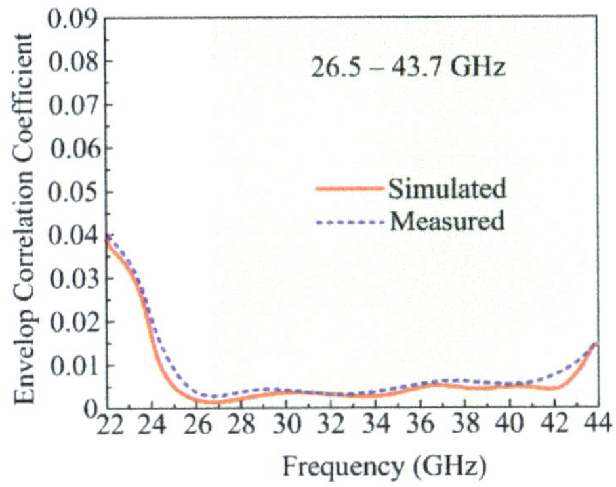

Figure 14. ECC of the proposed MIMO antenna.

In the above equation, *i* and *j* are antenna elements and *N* represents number of antennas placed in MIMO systems.

3.2.5. Diversity Gain

By using a MIMO antenna system, there are transmission power losses that may be examined by looking at diversity gain (DG). In the best-case situation, the value DG = 10 dB; however, in real-world scenarios, a value near 10 dB is taken into account. According to Figure 15, the suggested ultra-wideband MIMO antenna provides DG of about 9.99 dB over an operational bandwidth of 26.5–42.7 GHz. Equation (5) provided in [45,46] may be used to quantitatively compute the DG of a MIMO antenna system.

$$DG = 10\sqrt{1 - |ECC|^2} \quad (5)$$

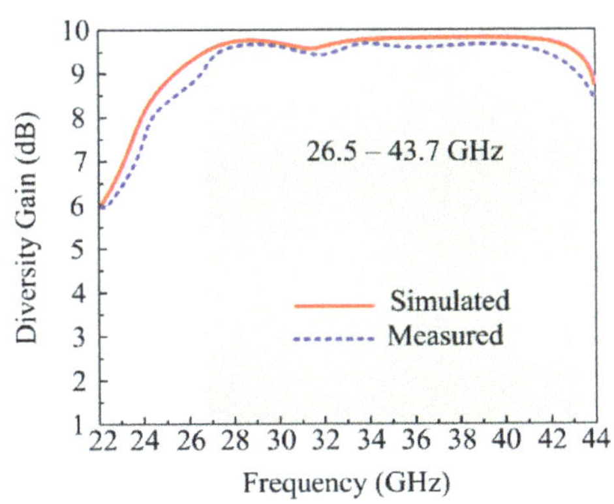

Figure 15. Comparison among predicted and measured DG.

3.2.6. Channel Capacity Loss

Channel capacity losses also happen as a result of correlation losses in MIMO antennas. It is a crucial component of every MIMO antenna. The value of CCL is observed for the proposed ultra-wideband MIMO antenna at about 0.01 bps/Hz over the operational bandwidth of 26.5–42.7 GHz, as shown in Figure 16. The value of CCL in an ideal situation is in the vicinity of 0.5 bps/Hz. The Formula (6) to compute CCL is shown below [45,46].

$$C_{Loss} = -log_2 \det\left(\propto^R\right) \quad (6)$$

where

$$\propto^R = \begin{bmatrix} \propto_{11} & \propto_{12} \\ \propto_{21} & \propto_{22} \end{bmatrix}$$

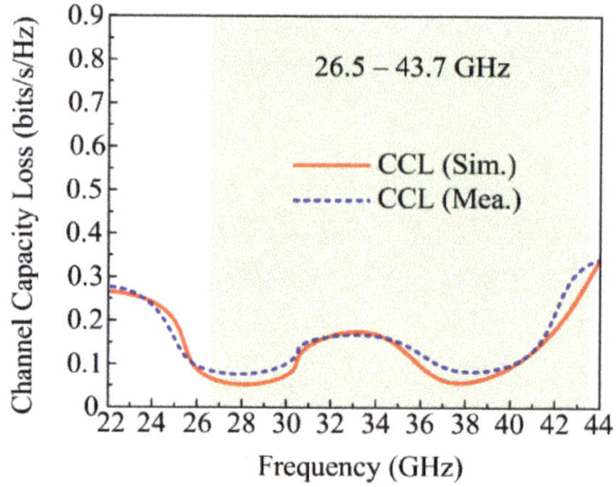

Figure 16. CCL of the proposed work.

3.2.7. Mean Effective Gain

By examining the mean effective gain, the received power of a MIMO antenna system in a fading environment may be examined (MEG). The ideal instance would have an MEG value of less than 3 dBi. According to Figure 17, the suggested ultra-wideband MIMO antenna gives MEG −8 dBi at an operating bandwidth of 26.5–42.7 GHz. Moreover, the following Equation (7), provided in [45,46], may be used to compute the MEG of a MIMO antenna:

$$MEG_i = \frac{P_{rec}}{P_{inc}} = \oint \left[\frac{XPR.\ G_{\theta i}(\Omega) + G_i(\Omega).P(\Omega)}{1 + XPR} \right] d\Omega \tag{7}$$

where $P_{\varnothing}$ represents an angular density function of incident power and XPR is the cross-polarization power ratio.

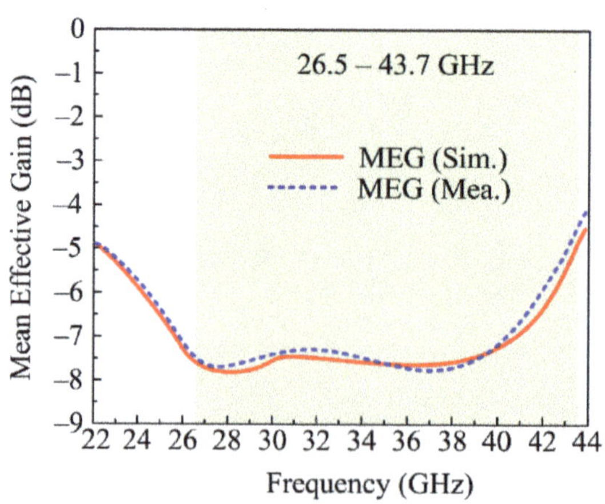

Figure 17. MEG of the proposed MIMO antenna.

3.3. Performance Comparison

Table 1 provides a comparison of the proposed ultra-wideband antenna operating over 26.5–43.7 GHz with the antenna already published in literature. The comparison in terms of size, number of ports, operational bandwidth, peak gain, ECC, and minimum/maximum isolation is performed. It can be seen from the table that antenna has compact size, offers wideband and high gain, along with good and acceptable values of ECC and isolation. This comparison also makes the proposed antenna a good applicant for future 5G devices having a compact size and operating over wideband and high gain.

Table 1. Comparison between suggested and literature works operating over same frequency.

Ref	Dimensions (mm × mm × mm)	Ports	Bandwidth (GHz)	Operating Frequency (GHz)	Peak Gain (dBi)	ECC	Mini. Isolation (dB)	Max. Isolation (dB)	MIMO Antenna Type
[18]	115 × 60 × 0.76	5	27.5–28.7	28.3	5	0.056	−30	−13	Monopole antenna
[19]	50.8 × 12.5 × 0.8	4	26–36	27	5.23	–	−45	−22	CPW-fed patch antenna
[20]	19.25 × 26 × 0.79	4	27–30.5	28.5	7.58	0.001	−35	−12	EBG-based antenna
[23]	24 × 20 × 1.85	4	33–44.1	38	4.56	0.1	−32	−16	Patch antenna over transparent substrate
[24]	53 × 20 × 0.203	2	22–50	36/45	15	0.12	−40	−20	Slot array antenna
[25]	110 × 55 × 1.6	6	27.7–28.7	28	5.13	0.005	−55	−22	Air-filled slotted loop (AFSL) antenna
[26]	12.5 × 12.5 × 0.8	4	33–36	35	6	0.02	−33	−23	Hexagonal patch antenna
[27]	60 × 100 × 0.965	2	27.6–28.3	28	4.5	0.134	−30	−17	Modified monopole antenna
[29]	48 × 31 × 0.254	4	26–31	28	10	0.0015	−38	−21	Patch antenna loaded with array of meatmaterials
[30]	80 × 80 × 1.57	4	23–40	30	12	0.0014	−40	−20	Arc-shaped patch antenna
[31]	20.5 × 12 × 0.79	2	25.5–30	28	8.75	–	−40	−30	E-shaped patch antenna
Prop. Work	27 × 27 × 1.52	4	26.5–43.7	30/38	8.4	0.001	−42	−30	Stub loaded monopole antenna

4. Conclusions

This article presents the design and validation of a wideband antenna for Ka-band 5G applications. The initial design is comprised of a rectangular printed antenna, whose performance is enhanced using an additional circular patch along with the loading of two open-ended stubs. The resultant antenna offers wideband ranges from 26.5 GHz to 43.7 GHz having a peak gain of >6 dBi in the operational band. The unit element had a compact size of 10 × 15 mm^2. Furthermore, to meet the requirements of the modern-day devices, a MIMO antenna is constructed using elements. The consecutive elements are placed orthogonal to achieve a low mutual coupling. Moreover, the presence of the open-ended stubs further helps to lower mutual coupling. The edge-to-edge space between consecutive elements is 4 mm and the maximum mutual coupling is less than −30 dB. Other performance parameters of the MIMO antenna are also studied, which show acceptable values in all terms. At last, the performance comparison is performed with recently reported works and the proposed work overperforms all the works by offering wideband, compact size, high gain, and low mutual coupling, making the proposed work a potential candidate for ka-band applications. The future extension of the proposed work is to deploy it for massive MIMO applications requiring the low mutual coupling among the element placed in close range.

Author Contributions: Conceptualization, methodology, software, validation, S.A.H., F.T., M.S.A. and I.H.; formal analysis, investigation, resources, data curation, R.M.G., M.F.A.S. and A.L.; writing—original draft preparation, S.A.H., I.H. and F.T.; writing—review and editing, M.S.A., R.M.G., M.F.A.S. and A.L.; visualization, R.M.G.; supervision, A.L.; project administration, funding acquisition, R.M.G. and A.L. All authors have read and agreed to the published version of the manuscript.

Funding: Princess Nourah bint Abdulrahman University Researchers Supporting Project number (PNURSP2023R138), Princess Nourah bint Abdulrahman University, Riyadh, Saudi Arabia.

Institutional Review Board Statement: Not applicable.

Informed Consent Statement: Not applicable.

Data Availability Statement: All data is included in the study.

Acknowledgments: We acknowledge the support from Princess Nourah bint Abdulrahman University Researchers Supporting Project number (PNURSP2023R138), Princess Nourah bint Abdulrahman University, Riyadh, Saudi Arabia.

Conflicts of Interest: The authors declare no conflict of interest.

References

1. Rappaport, T.S.; Sun, S.; Mayzus, R.; Zhao, H.; Azar, Y.; Wang, K.; Wong, G.N.; Schulz, J.K.; Samimi, M.; Gutierrez, F. Millimeter wave mobile communications for 5G cellular: It will work! *IEEE Access* **2013**, *1*, 335–349. [CrossRef]
2. Hussain, M.; Awan, W.A.; Alzaidi, M.S.; Hussain, N.; Ali, E.M.; Falcone, F. Metamaterials and their application in the performance enhancement of reconfigurable antennas: A review. *Micromachines* **2023**, *14*, 349. [CrossRef] [PubMed]
3. Kulkarni, J.; Sim, C.Y.D.; Gangwar, R.K.; Anguera, J. Broadband and compact circularly polarized MIMO antenna with concentric rings and oval slots for 5G application. *IEEE Access* **2022**, *10*, 29925–29936. [CrossRef]
4. Rosaline, I.; Kumar, A.; Upadhyay, P.; Murshed, A.H. Four element MIMO antenna systems with decoupling lines for high-speed 5G wireless data communication. *Int. J. Antennas Propag.* **2022**, *2022*, e9078929. [CrossRef]
5. Hei, Y.Q.; He, J.G.; Li, W.T. Wideband Decoupled 8-Element MIMO Antenna for 5G Mobile Terminal Applications. *IEEE Antennas Wirel. Propag. Lett.* **2021**, *20*, 1448–1452. [CrossRef]
6. Hussain, M.; Awan, W.A.; Ali, E.M.; Alzaidi, M.S.; Alsharef, M.; Elkamchouchi, D.H.; Alzahrani, A.; Fathy Abo Sree, M. Isolation improvement of parasitic element-loaded dual-band MIMO antenna for mm-Wave applications. *Micromachines* **2022**, *13*, 1918. [CrossRef]
7. Ali, E.M.; Awan, W.A.; Naqvi, S.I.; Alzaidi, M.S.; Alzahrani, A.; Elkamchouchi, D.H.; Falcone, F.; Alharbi, T.E.A. A low-profile antenna for on-body and off-body applications in the lower and upper ISM and WLAN bands. *Sensors* **2023**, *23*, 709. [CrossRef]
8. Tang, H.; Bulger, C.J.; Rovere, T.; Zheng, B.; An, S.; Li, H.; Dong, Y.; Haerinia, M.; Fowler, C.; Gonya, S.; et al. A Low-Profile Flexible Dual-Band Antenna with Quasi-Isotropic Radiation Patterns for MIMO System on UAVs. *IEEE Antennas Wirel. Propag. Lett.* **2022**, *22*, 49–53. [CrossRef]
9. Affandi, A.; Azim, R.; Alam, M.M.; Islam, M.T. A low-profile wideband antenna for WWAN/LTE applications. *Electronics* **2020**, *9*, 393. [CrossRef]
10. Zaidi, A.; Awan, W.A.; Ghaffar, A.; Alzaidi, M.S.; Alsharef, M.; Elkamchouchi, D.H.; Ghoneim, S.S.M.; Alharbi, T.E.A. A low profile ultra-wideband antenna with reconfigurable notch band characteristics for smart electronic systems. *Micromachines* **2022**, *13*, 1803. [CrossRef]
11. Hussain, M.; Ali, E.M.; Awan, W.A.; Hussain, N.; Alibakhshikenari, M.; Virdee, B.S.; Falcone, F. Electronically reconfigurable and conformal triband antenna for wireless communications systems and portable devices. *PLoS ONE* **2022**, *17*, e0276922. [CrossRef] [PubMed]
12. Hussain, M.; Hussain, A.; Alibakhshikenari, M.; Falcone, F.; Limiti, E. A simple geometrical frequency reconfigurable Antenna with Miniaturized Dimensions for 24.8/28GHz 5G Applications. In Proceedings of the 2022 16th European Conference on Antennas and Propagation (EuCAP), Madrid, Spain, 27 March–1 April 2022; pp. 1–3.
13. Kamal, M.M.; Yang, S.; Kiani, S.H.; Sehrai, D.A.; Alibakhshikenari, M.; Abdullah, M.; Falcone, F.; Limiti, E.; Munir, M. A novel hook-shaped antenna operating at 28 GHz for future 5G mmwave applications. *Electronics* **2021**, *10*, 673. [CrossRef]
14. Zahra, H.; Awan, W.A.; Ali, W.A.E.; Hussain, N.; Abbas, S.M.; Mukhopadhyay, S. A 28 GHz Broadband Helical Inspired End-Fire Antenna and Its MIMO Configuration for 5G Pattern Diversity Applications. *Electronics* **2021**, *10*, 405. [CrossRef]
15. Ibrahim, A.A.; Zahra, H.; Dardeer, O.M.; Hussain, N.; Abbas, S.M.; Abdelghany, M.A. Slotted Antenna Array with Enhanced Radiation Characteristics for 5G 28 GHz Communications. *Electronics* **2022**, *11*, 2664. [CrossRef]
16. Ullah, H.; Abutarboush, H.F.; Rashid, A.; Tahir, F.A. A Compact Low-Profile Antenna for Millimeter-Wave 5G Mobile Phones. *Electronics* **2022**, *11*, 3256. [CrossRef]
17. Kumar, P.; Ali, T.; Kumar, O.P.; Vincent, S.; Kumar, P.; Nanjappa, Y.; Pathan, S. An Ultra-Compact 28 GHz Arc-Shaped Millimeter-Wave Antenna for 5G Application. *Micromachines* **2023**, *14*, 5. [CrossRef]
18. Ikram, M.; Sharawi, M.S.; Shamim, A. A novel very wideband integrated antenna system for 4G and 5G mm-wave applications. *Microw. Opt. Technol. Lett.* **2017**, *59*, 3082–3088. [CrossRef]
19. Jilani, S.F.; Abbasi, Q.H.; Imran, M.A.; Alomainy, A. Design and Analysis of Millimeter-Wave Antennas for the Fifth Generation Networks and Beyond. In *Wiley 5G Ref: The Essential 5G Reference Online*; Wiley Online Library: New York, NY, USA, 2019; pp. 1–21.
20. Tu, D.T.T.; Thang, N.G.; Ngoc, N.T.; Phuong, N.T.B.; Van Yem, V. 28/38 GHz dual-band MIMO antenna with low mutual coupling using novel round patch EBG cell for 5G applications. In Proceedings of the International Conference on Advanced Technologies for Communications (ATC), Quy Nhon, Vietnam, 18–20 October 2017; pp. 64–69.

21. Xing, H.; Wang, X.; Gao, Z.; An, X.; Zheng, H.-x.; Wang, M.; Li, E. Efficient Isolation of an MIMO Antenna Using Defected Ground Structure. *Electronics* **2020**, *9*, 1265. [CrossRef]
22. Abdullah, M.; Kiani, S.H.; Iqbal, A. Eight element multiple-input multiple-output (MIMO) antenna for 5G mobile applications. *IEEE Access* **2019**, *7*, 134488–134495. [CrossRef]
23. Desai, A.; Bui, C.D.; Patel, J.; Upadhyaya, T.; Byun, G.; Nguyen, T.K. Compact wideband four element optically transparent MIMO antenna for mm-wave 5G applications. *IEEE Access* **2020**, *8*, 194206–194217. [CrossRef]
24. Saad, A.A.R.; Mohamed, H.A. Printed millimeter-wave MIMO-based slot antenna arrays for 5G networks. *AEU Int. J. Electron. Commun.* **2019**, *99*, 59–69. [CrossRef]
25. Marzouk, H.M.; Ahmed, M.I.; Shaalan, A.A. *A Novel Dual-Band 28/38 GHz AFSL MIMO Antenna for 5G Smartphone Applications*; Journal of Physics: Conference Series; IOP Publishing: Bristol, UK, 2020; Volume 1447.
26. Elfergani, I.; Rodriguez, J.; Iqbal, A.; Sajedin, M.; Zebiri, C.; AbdAlhameed, R.A. Compact millimeter-wave MIMO antenna for 5G applications. In Proceedings of the 14th European Conference on Antennas and Propagation (EuCAP), Copenhagen, Denmark, 15–20 March 2020; pp. 1–5.
27. Alreshaid, A.T.; Hussain, R.; Podilchak, S.K.; Sharawi, M.S. A dual-element MIMO antenna system with a mm-wave antenna array. In Proceedings of the 2016 10th European conference on antennas and propagation (EuCAP), Davos, Switzerland, 10–15 April 2016; pp. 1–4.
28. Arabi, O.; See, C.H.; Ullah, A.; Ali, N.; Liu, B.; Abd-Alhameed, R.; McEwan, N.J.; Excell, P.S. Compact Wideband MIMO Diversity Antenna for Mobile Applications Using Multi-Layered Structure. *Electronics* **2020**, *9*, 1307. [CrossRef]
29. Wani, Z.; Abegaonkar, M.P.; Koul, S.K. A 28-GHz antenna for 5G MIMO applications. *Prog. Electromagn. Res. Lett.* **2018**, *78*, 73–79. [CrossRef]
30. Sehrai, D.A.; Abdullah, M.; Altaf, A.; Kiani, S.H.; Muhammad, F.; Tufail, M.; Irfan, M.; Glowacz, A.; Rahman, S. A novel high gain wideband MIMO antenna for 5G millimeter wave applications. *Electronics* **2020**, *9*, 1031. [CrossRef]
31. Taher, F.; Hamadi, H.A.; Alzaidi, M.S.; Alhumyani, H.; Elkamchouchi, D.H.; Elkamshoushy, Y.H.; Haweel, M.T.; Sree, M.F.A.; Fatah, S.Y.A. Design and analysis of circular polarized two-port MIMO antennas with various antenna element orientations. *Micromachines* **2023**, *14*, 380. [CrossRef]
32. Dayo, Z.A.; Aamir, M.; Rahman, Z.; Khoso, I.A.; Lodro, M.M.; Dayo, S.A.; Soothar, P.; Pathan, M.S.; Al-Gburi, A.J.A.; Memon, A.A.; et al. A Novel Low-Cost Compact High-Performance Flower-Shaped Radiator Design for Modern Smartphone Applications. *Micromachines* **2023**, *14*, 463. [CrossRef] [PubMed]
33. Saeidi, T.; Al-Gburi, A.J.A.; Karamzadeh, S. A Miniaturized Full-Ground Dual-Band MIMO Spiral Button Wearable Antenna for 5G and Sub-6 GHz Communications. *Sensors* **2023**, *23*, 1997. [CrossRef]
34. Khan, A.; He, Y.; He, Z.; Chen, Z.N. A Compact Quadruple-Band Circular Polarized MIMO Antenna With Low Mutual Coupling. *IEEE Trans. Circuits Syst. II Express Briefs* **2023**, *70*, 501–505. [CrossRef]
35. Liu, J.; Liu, H.; Dou, X.; Tang, Y.; Zhang, C.; Wang, L.; Tang, R.; Yin, Y. A low profile, dual-band, dual-polarized patch antenna with antenna-filter functions and its application in MIMO systems. *IEEE Access* **2021**, *9*, 101164–101171. [CrossRef]
36. Khan, A.; He, Y.; Chen, Z.N. An Eight-Port Circularly Polarized Wideband MIMO Antenna Based on a Metamaterial-Inspired Element for 5G mmWave Applications. *IEEE Antennas Wirel. Propag. Lett.* **2023**. [CrossRef]
37. Ali, W.A.E.; Ibrahim, R.A. Highly Compact 4 × 4 Flower-Shaped MIMO Antenna for Wideband Communications. *Appl. Sci.* **2023**, *13*, 3532. [CrossRef]
38. Wong, K.L.; Jian, M.F.; Li, W.Y. Low-profile wideband four-corner-fed square patch antenna for 5G MIMO mobile antenna application. *IEEE Antennas Wirel. Propag. Lett.* **2021**, *20*, 2554–2558. [CrossRef]
39. Chattha, H.T.; Ishfaq, M.K.; Khawaja, B.A.; Sharif, A.; Sheriff, N. Compact multiport MIMO antenna system for 5G IoT and cellular handheld applications. *IEEE Antennas Wirel. Propag. Lett.* **2021**, *20*, 2136–2140. [CrossRef]
40. Din, I.U.; Kiyani, A.; Naqvi, S.I.; Al-Gburi, A.J.A.; Abbas, S.M.; Ullah, S. A Low-Cost Wideband MIMO Antenna for IoT Applications. In Proceedings of the 2022 IEEE International Symposium on Antennas and Propagation and USNC-URSI Radio Science Meeting (AP-S/URSI), Denver, CO, USA, 10–15 July 2022; pp. 2064–2065. [CrossRef]
41. Jehangir, S.S.; Sharawi, M.S. A compact single-layer four-port orthogonally polarized Yagi-like MIMO antenna system. *IEEE Antennas Wirel. Propag. Lett.* **2020**, *68*, 6372–6377. [CrossRef]
42. Hussain, M.; Mousa Ali, E.; Jarchavi, S.M.R.; Zaidi, A.; Najam, A.I.; Alotaibi, A.A.; Althobaiti, A.; Ghoneim, S.S.M. Design and characterization of compact broadband antenna and its MIMO configuration for 28 GHz 5G applications. *Electronics* **2022**, *11*, 523. [CrossRef]
43. Dayo, Z.A.; Aamir, M.; Dayo, S.A.; Khoso, I.A.; Soothar, P.; Sahito, F.; Zheng, T.; Hu, Z.; Guan, Y. A novel compact broadband and radiation efficient antenna design for medical IoT healthcare system. *Math. Biosci. Eng.* **2022**, *19*, 3909–3927. [CrossRef]
44. Dama, Y.A.S.; Abd-Alhameed, R.A.; Jones, S.M.R.; Zhou, D.; McEwan, N.J.; Child, M.B.; Excell, P.S. An envelope correlation formula for (N, N) MIMO antenna arrays using input scattering parameters, and including power losses. *Int. J. Antennas Propag.* **2011**, *2011*, 421691. [CrossRef]

45. Hussain, N.; Awan, W.A.; Ali, W.; Naqvi, S.I.; Zaidi, A.; Le, T.T. Compact wideband patch antenna and its MIMO configuration for 28 GHz applications. *AEU Int. J. Electron. Commun.* **2021**, *132*, 153612. [CrossRef]
46. Cai, J.; Zhang, J.; Xi, S.; Huang, J.; Liu, G. A Wideband Eight-Element Antenna with High Isolation for 5G New-Radio Applications. *Appl. Sci.* **2023**, *13*, 137. [CrossRef]

applied sciences

MDPI

Communication

A Compact Size Antenna for Extended UWB with WLAN Notch Band Stub

Syed Naheel Raza Rizvi [1], Wahaj Abbas Awan [1], Domin Choi [1], Niamat Hussain [2], Seong Gyoon Park [3] and Nam Kim [1,*]

1 Department of Information and Communication Engineering, Chungbuk National University, Cheongju 28644, Republic of Korea
2 Department of Smart Device Engineering, Sejong University, Seoul 05006, Republic of Korea
3 Department of Information and Communication Engineering, Kongju National University, Gongju-si 32588, Republic of Korea
* Correspondence: namkim@chungbuk.ac.kr

Abstract: An ultra-wideband (UWB), geometrically simple, compact, and high-gain antenna with a WLAN notch band is presented for future wireless devices. The antenna is printed on the top side of the Rogers RT/Duroid 5880 substrate and has a small dimension of 10 mm × 15 mm × 0.254 mm. The primary radiator of the proposed coplanar waveguide-fed monopole antenna is comprised of a rectangular-shaped structure initially modified using a slot, and its bandwidth is further enhanced by loading a Y-shaped radiator. As a result, the antenna offers a –10 dB impedance matching bandwidth of 11.55 GHz ranging from 3–14.55 GHz, covering globally allocated C-, S-, and X-band applications. Afterward, another rectangular stub is loaded in the structure to mitigate the WLAN band from the UWB spectrum, and the final antenna offers a notched band spanning from 4.59 to 5.82 GHz. Moreover, to validate the simulated results, a hardware prototype is built and measured, which exhibits good agreement with the simulated results. Furthermore, the proposed work is compared to state-of-the-art antennas for similar applications to demonstrate its design significance, as it has a compact size, wider bandwidth, and stable gain characteristics.

Keywords: UWB antenna; compact antenna; notch-band antenna; WLAN band rejection

Citation: Rizvi, S.N.R.; Awan, W.A.; Choi, D.; Hussain, N.; Park, S.G.; Kim, N. A Compact Size Antenna for Extended UWB with WLAN Notch Band Stub. *Appl. Sci.* **2023**, *13*, 4271. https://doi.org/10.3390/app13074271

Academic Editor: Naser Ojaroudi Parchin

Received: 16 February 2023
Revised: 23 March 2023
Accepted: 26 March 2023
Published: 28 March 2023

1. Introduction

The recent evolution in wireless communication systems has urged us to revise the requirements for designing communicating gadgets. The high data rate transfer with low latency and low power consumption in ultra-wideband (UWB) technology has opened doors for many applications in modern society [1,2]. The UWB technology is not limited to satellite communication or military purposes, but it also finds many applications, ranging from biomedical imaging to wireless sensor networks [3–5]. Thus, having an eye on the advantages of the UWB spectrum, a number of antennas have been proposed in the literature [6–12].

The UWB antennas presented in [6–8] have larger physical dimensions and do not cover the UWB spectrum (3.1 GHz to 10.6 GHz) allocated globally. Thus, it results in limiting their application to modern-day compact electronics. On the other hand, the compact-sized antennas designed in [9,10] have drawbacks of structural complexity, low gain, and unstable performance, while others have the disadvantage of limited bandwidth [11,12]. Moreover, it is necessary to design notch band UWB antennas due to the existence of sub-bands inside the UWB spectrum, like Wireless Local Area Network (WLAN), Wireless Fidelity (Wi-Fi), Worldwide Interoperability for Microwave Access (WiMAX), and Industrial, Scientific, and Medical (ISM) bands. Wi-Fi interference with UWB spectrum may also cause several co-existence issues like packet capturing and degraded performance [13,14]. It is noted that the UWB antennas proposed in [6–12] do not offer on-demand notch band functionality. Therefore, a lot of research has been done in designing new antennas for UWB spectrum with notching characteristics. Since the 5 GHz band spectrum is widely used for a number of applications, there is a dire need to mitigate this band from the UWB region to minimize the interference with the UWB communication systems. Therefore, various UWB antenna systems in the literature [15] are focusing on efficient antenna systems offering low interference. Hence, to meet the requirement of present day UWB applications, UWB antennas should offer notch band functionality without sacrificing overall antenna performance in terms of bandwidth, gain, and stable radiation pattern [16,17].

Several UWB antennas with WLAN notch band characteristics that have the advantages of compact electrical size have been widely investigated in the literature [18–30]. For instance, a circular patch antenna that has a balanced notch band feature with a common mode suppression is presented in [18]. A stub-loaded circular patch-based antenna was designed to achieve UWB mode. Afterward, a bandpass filter was designed in such a way that the resonator is judiciously fed, resulting in a common notch band behavior. However, due to the utilization of a filter, the overall size of the antenna gets increased, and the optimized antenna offers a physical size of 29.38 mm × 28.25 mm. Likewise, a circular-shaped monopole antenna is designed in [26] to achieve wide operational mode ranges from 2.8–12 GHz, while an enhanced mushroom-like modified Electromagnetic Band Gap (EBG) structure is utilized to notch the desired band. However, the reported antenna has several drawbacks, including structural complexity and a larger physical dimension of 40 mm × 50 mm.

A rectangular-shaped quarter-wave monopole antenna was modified using two consecutive iterations of slots to design a modified W-shaped fractal antenna for UWB applications [27]. Additionally, a pair of two inverted L-shaped open-ended stubs were loaded with a ground plane to achieve notch band functionality. The resulting antenna operates from 3–12 GHz, a notched band of 5.2–5.8 GHz, and a set back of a large overall size of 31 mm × 27 mm. On the other hand, a multi-layered Artificial Magnetic Conductor (AMC) loaded UWB antenna is presented in [28]. At first, a rectangular monopole was modified using truncated corners to widen the bandwidth, then two modified U-shaped slots were etched to achieve a notch band, and finally, an AMC was loaded to enhance the performance of the antenna. Having the advantage of high gain and two notch bands, the antenna still has some disadvantages, including its limited bandwidth of 4.8–10 GHz and its large physical size of 33 mm × 33 mm, with an overall critical height of 7 mm.

Last but not least, a pac-man-shaped printed UWB antenna was designed to achieve two distinct notch bands [29]. However, the notch band does not have the potential for practical usage due to poor comparison between simulated and measured results.

Winding up the whole discussion, there is still a need for a UWB antenna that has notch band functionality without degrading the other design constraints like compact size, wideband, geometrically simple, high gain, and stable radiation patterns. Thus, an antenna having a wide operational bandwidth with a WLAN notch band characteristic is proposed in this paper. The organization of the remaining manuscript is as under. The design methodology of the proposed UWB antenna and respective results are presented in Section 2, while the performance parameters are analyzed in Section 3. Furthermore, Section 4 contains the comparison of the proposed work with the state-of-the-art, while the manuscript is concluded in Section 5, followed by references.

2. Design Methodology of the UWB Antenna

2.1. Geometry of the Proposed Antenna

The top and side perspectives of the presented UWB antenna are depicted in Figure 1. The antenna has a small footprint of only 15 mm × 10 mm (L × W). The antenna geometry has a relative electrical permittivity and a loss tangent of 2.2 and 0.002, respectively, with a thickness of 0.254 mm, and is etched on the Rogers 5880/Duroid substrate. The proposed UWB antenna consists of a Coplaner Waveguide (CPW) feedline and a rectangular patch with several stubs and slots. An additional stub is added to the modified rectangular patch radiator to enhance impedance bandwidth and get WLAN spectrum rejection. The proposed antenna is of the CPW feedline which allows for the inherent advantages of broadband performance, decreased dispersion, and ease of manufacture.

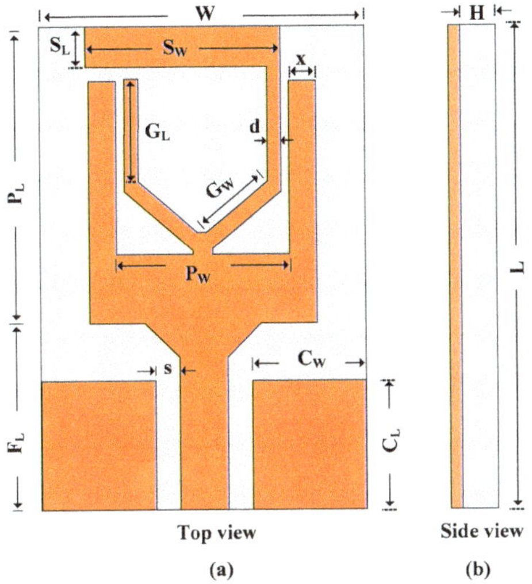

Figure 1. The geometry of the proposed WLAN Notch UWB antenna. (**a**) Top view, (**b**) side view.

The following are the optimal dimensions of the proposed notch band UWB antenna: W = 10, L= 15, H = 0.254, P_l = 9.2, Pw = 5.4, C_L = 4, C_W = 3.5, s = 0.75, x = 0.8, S_L = 1.25, S_W = 6, d = 0.4, G_L = 3.19, and G_W = 2.65 (All dimensions are in millimeters).

2.2. Antenna Design Steps

The design methodology steps and their respective s-parameters are shown in Figure 2. The proposed work consists of two major phases. Phase 1 consists of designing a compact size UWB antenna, which is further utilized in phase 2 to achieve UWB behavior with a WLAN notch band spectrum.

Initially, a conventional quarter-wave monopole antenna that has a rectangular radiator was designed. The commercially available CST Microwave Studio software was used to perform the antenna simulations. The antenna is CPW-fed, a potential feeding technique to achieve compact size, easy design, and broadband impedance matching. Moreover, its uni-planar structure also eases its integration with electronic circuitry. The respective length of the rectangular radiator for the desire frequency is estimated using the following equation provided in [30]:

$$L_{fo} = \frac{c}{4f_o\sqrt{e_{ff}}} \tag{1}$$

where c refers to the speed of light in the vacuum and f_o is the central resonating frequency of 6.5 GHz. Furthermore, e_{ff} is the constant representing the effective dielectric constant whose value can be approximated using Equation (2) [31]:

$$e_{ff} \approx \frac{e_r + 1}{2} \tag{2}$$

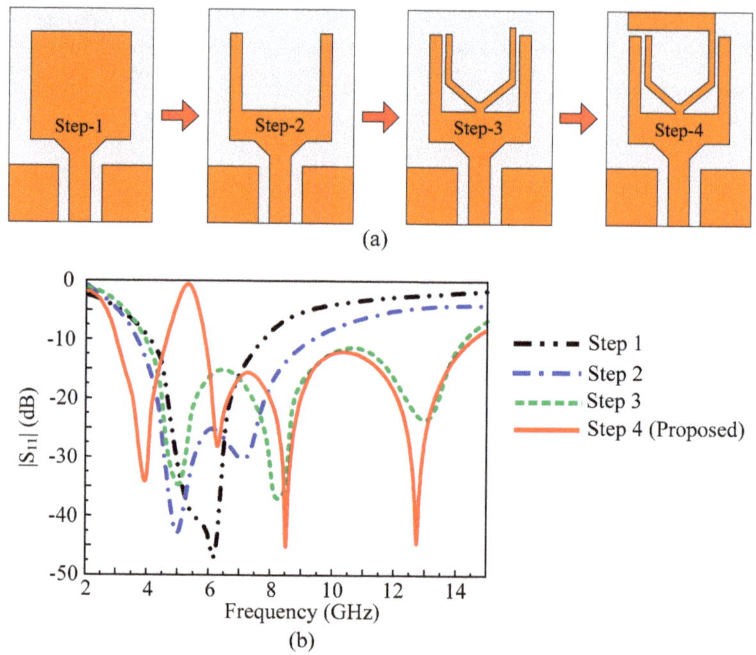

Figure 2. Proposed antenna's (**a**) various design steps. (**b**) $|S_{11}|$ responses.

After a slight optimization in parameters owing to the losses in the substrate material, the antenna offers a broad bandwidth of 5.4–8.2 GHz, as depicted in Figure 2b. Afterward, a slot was etched from the radiator, as shown in Figure 2a. This helps to achieve a further wider bandwidth without affecting the overall size of the antenna. The slot results in a Y-shaped radiator, which is also a potential structure to achieve wide impedance matching to cover UWB frequency band. The working principle of the Y-shaped radiator is explained in [32]. As shown in Figure 2b, the antenna now has a $|S_{11}|$ -10 dB impedance bandwidth of 4.8–9.4 GHz.

In the final step in designing the proposed UWB antenna, a rectangular stub on a Y-shaped radiator is introduced, as shown in Figure 2a. The antenna operates between 5 and 14 GHz, covering more than 85% of the UWB spectrum and extended UWB spectrums of 11 and 13 GHz, as recommended by the International Telecommunication Union (ITU) [33]. Finally, to avoid the interference of the UWB antenna with sub-bands lying inside the UWB spectrum, a rectangular stub was loaded on the top side of the radiator, which notches the globally allocated WLAN band spectrum of 5.25–5.85 GHz [34]. This specific frequency band spectrum covers 4.9 GHz and 5 GHz Wireless LAN (WLAN) accessed using IEEE 802.11 communication protocols. This spectrum has a number of frequency channels, each having a dynamic bandwidth of 10 MHz, 20 MHz, 40 MHz, 80 MHz, and 160 MHz, respectively. The Radio Standard Specification (RSS) provides a transmitter output power line and equivalent isotropically radiated power (EIRP) requirements, limiting the maximum EIRP peak value to 4 Watts. Therefore, the wireless signals from WLAN spectrum have the ability to interfere with the UWB spectrum. Furthermore, the notched band channels also have weather radars operating in these bands. Thus, the most congested part of the WLAN band spectrum 4.9 GHz (802.11j) and 5 GHz (802.11a/h/j) from channel 7 to channel 161 is attenuated for low interference with the UWB spectrum. Final optimization was done to achieve the maximum possible bandwidth while keeping the notch band constant. The proposed antenna offers UWB ranges from 3–14.55 GHz, having a notched band of 4.59–5.82 GHz, as depicted in Figure 2b.

2.3. Parametric Analysis

As stated earlier, the additional stub loaded at the top edge of the UWB radiator is responsible for the notch band behavior. Thus, to further demonstrate the findings, a parametric analysis was done by varying the length of the stub (S_W), and the corresponding $|S_{11}|$ results are depicted in Figure 3a. It can be observed from Figure 3a that when the optimized length of the stub is increased, the notch band shifts towards the left side, while the bandwidth of the notched area remains unchanged. Similarly, when the length of the stub is reduced, the notch band shifts toward the left side. When the electrical dimensions of an antenna are reduced, it results in a frequency response shift towards the right end.

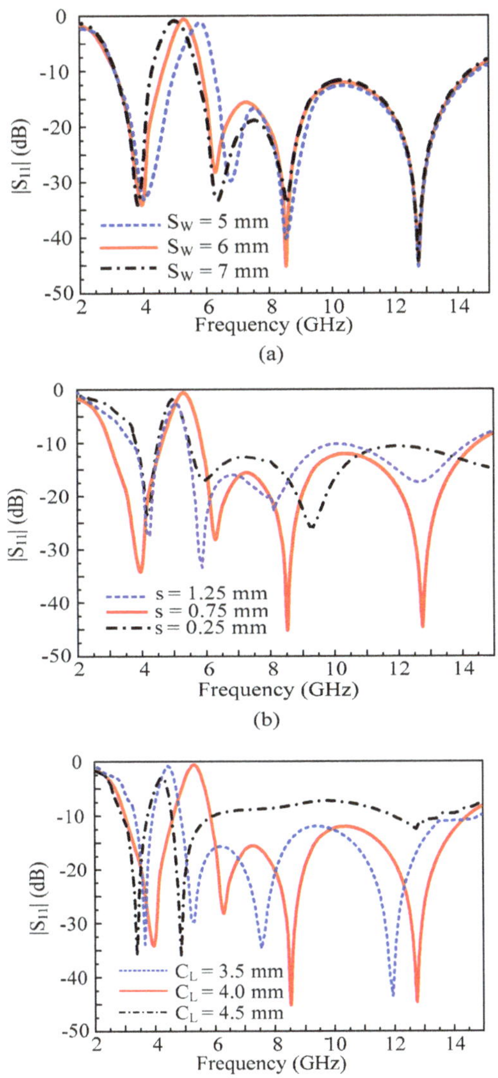

Figure 3. $|S_{11}|$ of the antenna for different values of (**a**) S_W, (**b**) s, (**c**) C_L.

Figure 3b,c demonstrate the $|S_{11}|$ results with respect to the variations in gap (s) and the length C_L. The characteristic impedance Zo of the coplanar waveguide varies as its dimensions (s and C_L) are varied. When the gap length is decreased, the notch band is shifted towards the lower end of the spectrum. Furthermore, it also causes a decrease in the bandwidth of the antenna due to mismatching of the impedance. However, when the gap length is increased, the bandwidth of the antenna decreases, and the notch band is shifted towards the left side. Similarly, when the length C_L is increased from the optimized value, the notch band shifts towards the left side.

3. Results and Discussions

3.1. Hardware Prototype and Measurement Setup

To verify simulated results, the hardware prototype of the proposed compact antenna was fabricated using a commercially available substrate. Then, the return loss of the proposed work was measured using Vector Network Analyzer (VNA), while far-field parameters, including gain and radiation pattern, were measured using an electromagnetically isolated RF anechoic chamber. Figure 4 depicts the fabricated prototype along with the far-field measurement setup.

(a) (b)

Figure 4. Proposed extended UWB antenna's (**a**) prototype and (**b**) antenna under test.

3.2. Reflection Coefficient

A comparison among simulated and measured reflection coefficient of the antenna is presented in Figure 5. It can be seen clearly that the antenna offers an ultra-wideband spectrum of 3–14.32 GHz when measured, having a notched band ranging from 4.59–5.82 GHz. Moreover, a strong comparison between simulated and measured results validates the findings.

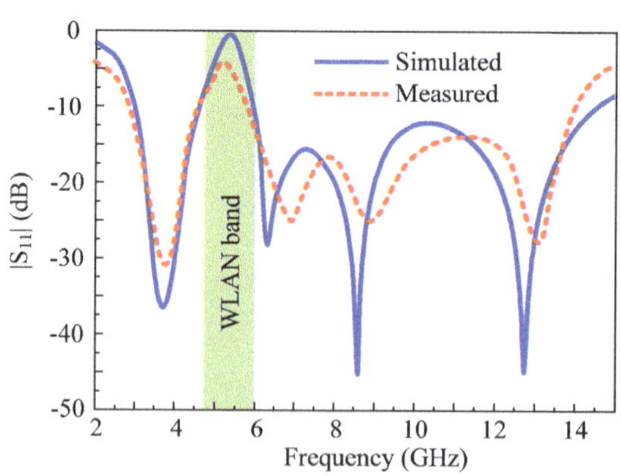

Figure 5. Measured and simulated $|S_{11}|$ of the proposed notch band extended UWB antenna.

3.3. Gain

The gain comparison of the proposed antenna for both simulated and measured values for the complete working region is depicted in Figure 6. The antenna has a simulated gain of more than 2 dBi in the resonating bandwidth, but it begins to decrease in the notch band, reaching a low of −3.9 dB at 5.5 GHz. The measured value of the gain also depicts similar results for both resonating and notch bandwidths. The antenna gain and far-field radiation characteristics are measured in an anechoic facility. A well-calibrated regular gain horn antenna was placed in the anechoic facility and utilized as a transmit antenna, whereas the prototype antenna acts as a receiver antenna. Low noise amplifiers were utilized for the provision of constant power reception. The antenna was rotated at a number of angles to measure the results at various configurations.

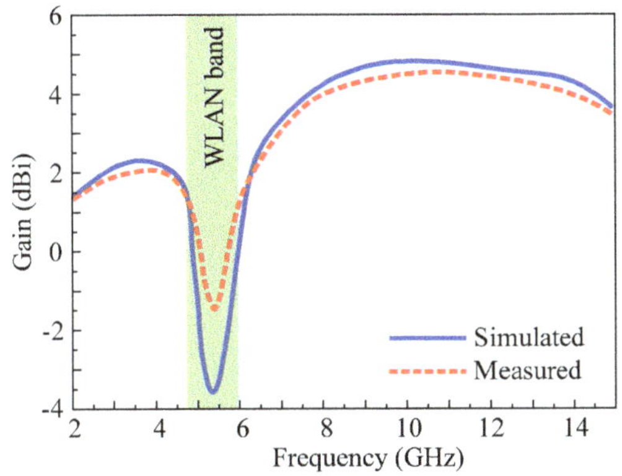

Figure 6. Simulated and measured gain of the proposed notch band extended UWB antenna.

3.4. Surface Current Distribution

To further understand the resonating behavior of the proposed antenna, Figure 7 illustrates surface current distribution at various resonating and notch band frequencies. It can be observed that at the lower resonance of 4 GHz, the current distribution is lower at the upper part of the radiator, which results in the generation of that frequency. Contrary to that, at the notch band, the current show maximum value around the upper stub and across it, while at the lower side, almost no current distribution is available. This causes the mitigation of the 5.5 GHz band. Furthermore, it can also be observed from Figure 7 that current is distributed across the radiator, which causes the generation of a higher resonance. Hence, it is also verified that the upper stub (rectangular stub) is responsible for mitigating the WLAN band from the UWB spectrum.

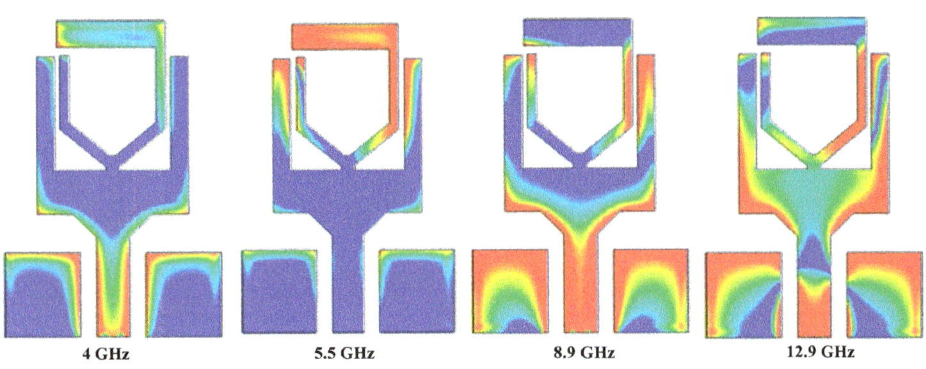

Figure 7. Current density of proposed notch band extended UWB antenna at various frequencies.

3.5. Radiation Pattern

The radiation pattern of the proposed UWB antenna with a notch band is presented in Figure 8. The antenna offers a nearly omni-directional radiation pattern in the principal

H-Plane for both resonating frequencies of 3.5 GHz and 6.5 GHz. On the other hand, for E-Plane, a dipole-like bidirectional radiation pattern was observed for 3.5 GHz, while a slightly tilted bidirectional radiation pattern was observed for 13 GHz, as depicted in Figure 8. In general, a strong agreement between simulated and measured results was observed for both resonating frequencies. It is worth noting that the antenna offers stable radiation patterns at all resonating frequencies, ensuring its wideband operation.

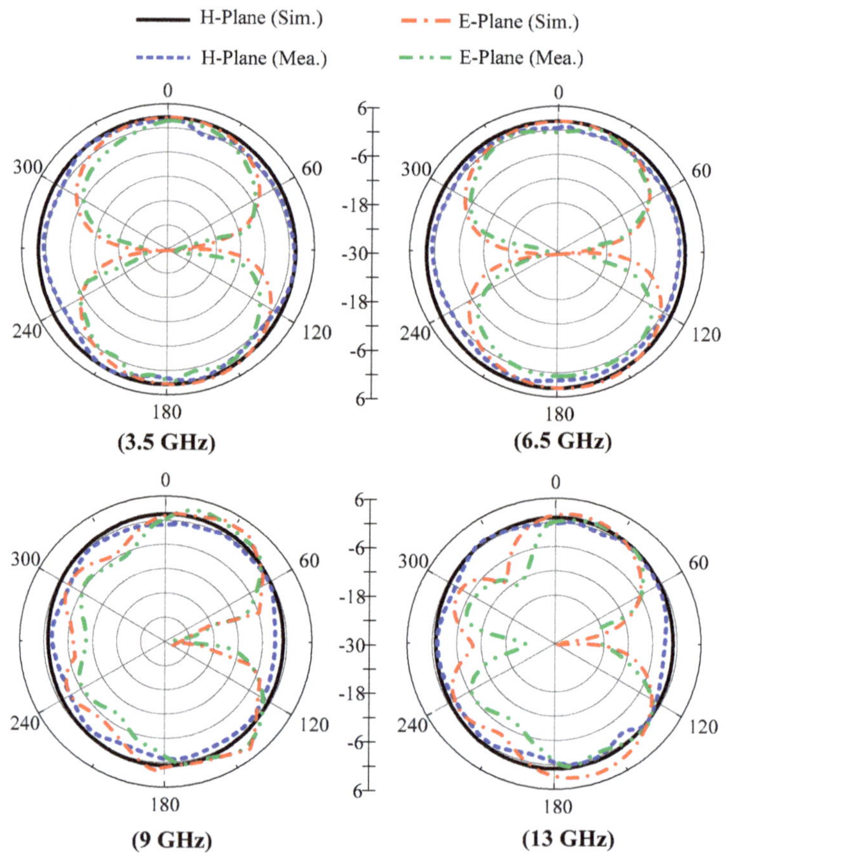

Figure 8. The radiation pattern of proposed notch band extended UWB antenna at various frequencies.

3.6. Group Delay

Group delay may be used to characterize the degree of distortion in the UWB antenna. In a UWB system, non-uniformity in group delay is desirable [31]. However, keeping the various losses with a small amount of non-uniformity is acceptable. Thus, two identical antennas are placed face-to-face and non-face-to-face at 30 cm to generate a far-field scenario. The group delay of the proposed notch band UWB is illustrated in Figure 9, where, except for the notch band, the group delay is entirely consistent. Except for the notch bands where the value is higher than >4 ns, the group delay varies less than 1 ns for the operational frequency range. Based on these results, the proposed antenna can be utilized for the UWB system, and there will be no distortion between the transmitting and receiving antennas.

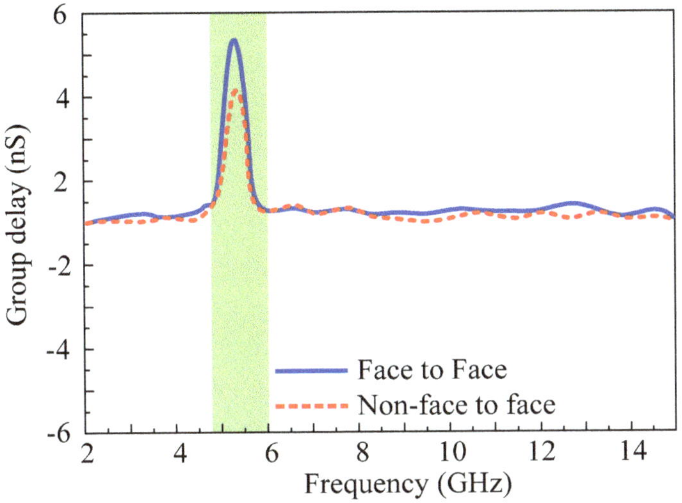

Figure 9. The group delay of the proposed notch band extended UWB antenna.

4. Comparison with State-of-the-Art UWB Antennas

Table 1 summarizes the proposed antenna's performance and its comparison with related work presented in the literature in recent years. It can be observed that the presented work offers a compact size as compared to all of the compared literature. The work reported in [15,17] offered dual-notch characteristics, but their size is twice as big as the proposed work. In addition, Refs. [17,18] offer a bit high peak gain value, but they suffer from structural complexity, along with the narrowband and big size as compared to the presented work. A semicircular-shaped printed antenna with two notch bands ranging from 3.3–4.2 GHz and 6.6–7.6 GHz was proposed in [19]. The lower notch band was realized using a modified dumbbell-shaped slot etched from the radiator, while the upper notch band was attained by employing a two-pair of capacitors, along with L-shaped slots carved from the ground plane. The antenna covers the complete UWB spectrum allocated globally. However, two pairs of capacitors along with narrow slots increase the structural complexity, while the physically large size of the antenna limits its applications for compact devices. In [20], a jug-shaped CPW antenna was presented. The antenna was fabricated on cheap FR4 substrate. However, the antenna does not offer any band rejection and may suffer from interference with WLAN spectrum. In [21], an EBG-backed monopole patch was presented. The antenna offers good notching characteristics. However, the antenna has a structurally large width of 25 mm. In [22], a metamaterial-backed monopole patch was presented, which offers a high gain. Although the presented design is a good candidate for high gain applications, the antenna offers no band-notched characteristics or interference rejection. Furthermore, an octagonal star-like monopole patch was presented in [23]. The antenna offers a fractional bandwidth of 120 percent. However, the antenna covers an area of 1275 mm^2, which is geometrically larger than the presented antenna in this manuscript. In [24], a unique geometry of an antenna was presented, offering a bandwidth of 3.581–14.1 GHz. The antenna has a peak moderate gain of 3.2 dBi. This antenna does not offer any band rejection capability and may not be suitable for applications where low interference is a vital requirement. In [25], a planer stub-loaded patch was presented. The antenna offers dual band-notched characteristics. However, the presented design is about 20% structurally larger than the presented design. Based upon the above-mentioned discussion, the proposed UWB antenna with WLAN notch band overperforms all the related work by offering a compact size, wideband, comparable peak gain, and low structural complexity.

Table 1. Performance comparison of the proposed antenna with existing UWB antennas in the literature.

Ref.	Dimensions ($\lambda_o \times \lambda_o \times \lambda_o$)	Frequency Range (GHz)	Rejection Band (GHz)	Peak Gain (dBi)	Design Methodology
[15]	0.3 × 0.31 × 0.02	2.9–14.5	3.1–3.6/4.9–6.1	No. Info.	Slot and SRR-loaded Antenna
[16]	0.21 × 0.21 × 0.02	3.04–10.87	5.03–5.94	4.2	Slot and SRR-loaded Antenna
[17]	0.29 × 0.33 × 0.08	2.88–12.67	3.43–3.85/5.26–6.01	4.6	Slot and SRR-loaded Antenna
[18]	0.32 × 0.30 × 0.005	2.95–10.75	5.01–6.19	5.2	Band Pass Filter-loaded Antenna
[19]	0.25 × 0.32 × 0.02	3.1–11.2	3.3–4.2/6.6–7.6	5.1	Capacitor-loaded Patch
[20]	0.255 × 0.222 × 0.02	3–11	None	4.1	Jug-shaped Monopole Patch
[21]	0.17 × 0.27 × 0.02	3.1–12.5	5–6	4.5	EBG-backed Monopole Patch
[22]	0.148 × 0.226	3.08–14.1	None	4.54	Metamaterial-backed Antenna
[23]	0.44 × 0.32 × 0.08	3.25–13	5.7–6.2	6.7	Octagonal Star-Like Patch
[24]	0.32 × 0.3 × 0.02	3.581–14.1	None	3.2	Slotted CPW-Fed Antenna
[25]	0.24 × 0.31 × 0.01	3.4–11.9	4.5–5.3/7.2–9	3.9	Stub-loaded Planer Patch
This Work	0.11 × 0.19 × 0.002	3–14.55	4.59–5.82	4.93	Stub-loaded Printed Antenna

5. Conclusions

A CPW-fed antenna with compact dimensions of 15 mm × 10 mm × 0.254 mm for WLAN notch band UWB application was presented. The antenna consists of simple geometry with a modified rectangular shaped radiator loaded with a Y-shaped radiator, improving impedance bandwidth ranges 3–14.55 GHz. Hereafter, an additional stub was loaded to achieve WLAN band rejection capability. The proposed antenna offers key features of compact dimensions, wide bandwidth, and moderate gain. This UWB antenna also has band rejection at 4.59–5.82 GHz, thereby reducing the interference with the WLAN band. The specific notched band channel is the most congested part of the WLAN band spectrum 4.9 GHz (802.11j) and 5 GHz (802.11a/h/j) from channel 7 to channel 161. This spectrum is attenuated for reducing the interference with the UWB spectrum. The antenna offers a stable radiation pattern with a gain of more than 2 dBi in the pass band, while for the notch band, a minimum gain of –3.9 dB was observed. The measurement results of the fabricated antenna are well-matched to the simulation results. Moreover, the presented antenna overperforms the literature work by offering a compact size, having a notched band, a simple geometrical configuration, a moderate gain, and an ultra-wide impedance bandwidth. Therefore, the presented antenna is a worthy candidate for utilization in UWB applications where WLAN band rejection is a vital requirement for operation.

Author Contributions: Conceptualization, methodology, software, validation, and formal analysis, S.N.R.R. and N.H.; investigation, resources and data curation, D.C., S.G.P. and N.K.; writing—original draft preparation, S.N.R.R. and W.A.A.; writing—review and editing, D.C., N.H., S.G.P. and N.K.; supervision, project administration and funding acquisition, N.K. All authors have read and agreed to the published version of the manuscript.

Funding: This work was supported by the Institute of Information & Communications Technology Planning & Evaluation (IITP) grant funded by the Korean government (MSIT) (No. 2022-0-01031, Development of measured EMF big data analysis and management platform).

Institutional Review Board Statement: Not applicable.

Informed Consent Statement: Not applicable.

Data Availability Statement: Not applicable.

Conflicts of Interest: The authors declare no conflict of interest.

References

1. Win, M.Z.; Dardari, D.; Molisch, A.F.; Wiesbeck, W.; Zhang, W.J. History and applications of UWB. *Proc. IEEE* **2009**, *97*, 198–204. [CrossRef]
2. Kumar, O.P.; Kumar, P.; Ali, T.; Kumar, P.; Vincent, S. Ultrawideband antennas: Growth and evolution. *Micromachines* **2021**, *13*, 60. [CrossRef]
3. Abbas, A.; Hussain, N.; Sufian, M.A.; Jung, J.; Park, S.M.; Kim, N. Isolation and gain improvement of a rectangular notch UWB-MIMO antenna. *Sensors* **2022**, *22*, 1460. [CrossRef] [PubMed]
4. Khan, M.S.; Naqvi, S.A.; Iftikhar, A.; Asif, S.M.; Fida, A.; Shubair, R.M. A WLAN band-notched compact four element UWB MIMO antenna. *Int. J. RF Microw. Comput. Aided Eng.* **2020**, *30*, 22282. [CrossRef]
5. Perli, B.R.; Avula, M.R. Design of wideband elliptical ring monopole antenna using characteristic mode analysis. *J. Electromagn. Eng. Sci.* **2021**, *21*, 299–306. [CrossRef]
6. Yeom, I.; Jung, Y.B.; Jung, C.W. Wide and dual-band MIMO antenna with omnidirectional and directional radiation patterns for indoor access points. *J. Electromag Eng. Sci.* **2019**, *19*, 20–30. [CrossRef]
7. Tangwachirapan, S.; Thaiwirot, W.; Akkaraekthalin, P. Design and analysis of antipodal vivaldi antennas for breast cancer detection. *Comput. Mater. Contin.* **2022**, *73*, 411–431.
8. Al-Gburi, A.J.A.; Ibrahim, I.B.M.; Zakaria, Z.; Ahmad, B.H.; Shairi, N.A.B.; Zeain, M.Y. High gain of uwb planar antenna utilising FSS reflector for UWB applications. *Comput. Mater. Contin.* **2022**, *70*, 1419–1436.
9. Jan, N.A.; Kiani, S.H.; Muhammad, F.; Sehrai, D.A.; Iqbal, A.; Tufail, M.; Kim, S. V-shaped monopole antenna with chichena itzia inspired defected ground structure for UWB applications. *Comput. Mater. Contin.* **2020**, *65*, 19–32.
10. Jan, N.A.; Kiani, S.H.; Sehrai, D.A.; Anjum, M.R.; Iqbal, A.; Abdullah, M.; Kim, S. Design of a compact monopole antenna for UWB applications. *Comput. Mater. Contin.* **2021**, *66*, 35–44.
11. Awan, W.A.; Choi, D.M.; Hussain, N.; Elfergani, I.; Park, S.G.; Kim, N. A frequency selective surface loaded UWB antenna for high gain applications. *Comput. Mater. Contin.* **2022**, *73*, 6169–6180.
12. Al-Gburi, A.J.A.; Zakaria, Z.; Palandoken, M.; Ibrahim, I.M.; Althuwayb, A.A.; Ahmad, S.; Al-barwi, S.S. Super compact UWB monopole antenna for small IoT devices. *Comput. Mater. Contin.* **2022**, *73*, 2785–2799.
13. Prasad, A.; Verma, S.S.; Dahiya, P.; Kumar, A. A case study on the monitor mode passive capturing of WLAN packets in an on-the-move setup. *IEEE Access.* **2021**, *9*, 152408–152420. [CrossRef]
14. Brunner, H.; Stocker, M.; Schuh, M.; Schub, M.; Boano, C.A.; Romer, K. Understanding and mitigating the impact of wi-fi 6e interference on Ultra-Wideband Communications and ranging. In Proceedings of the 2022 21st ACM/IEEE International Conference on Information Processing in Sensor Networks (IPSN), Milano, Italy, 4–6 May 2022; pp. 92–104.
15. Ramakrishna, C.; Kumar, G.S.; Reddy, P.C.S. Quadruple band-notched compact monopole UWB antenna for wireless applications. *J. Electromag Eng. Sci.* **2021**, *21*, 406–416. [CrossRef]
16. Ojaroudi, M.; Ojaroudi, N. Ultra-wideband small rectangular slot antenna with variable band-stop function. *IEEE Trans. Antennas Propag.* **2013**, *62*, 490–494. [CrossRef]
17. Jin, Y.; Tak, J.; Choi, J. Quadruple band-notched trapezoid UWB antenna with reduced gains in notch bands. *J. Electromag Eng. Sci.* **2016**, *16*, 35–43. [CrossRef]
18. Lee, C.H.; Wu, J.H.; Hsu, C.I.G.; Chan, H.L.; Chen, H.H. Balanced band-notched UWB filtering circular patch antenna with common-mode suppression. *IEEE Antennas Wirel. Propag. Lett.* **2017**, *16*, 2812–2815. [CrossRef]
19. Haider, A.; Rahman, M.; Ahmad, H.; Jahromi, M.N.; Niaz, M.T.; Kim, H.S. Frequency-agile WLAN notch UWB antenna for URLLC applications. *Comput. Mater. Contin.* **2021**, *67*, 2243–2254. [CrossRef]
20. Ahmad, S.; Ijaz, U.; Naseer, S.; Ghaffar, A.; Qasim, M.A.; Abrar, F.; Parchin, N.O.; See, C.H.; Abd-Alhameed, R. A jug-shaped CPW-fed ultra-wideband printed monopole antenna for wireless communications networks. *Appl. Sci.* **2022**, *12*, 821. [CrossRef]
21. Abbas, A.; Hussain, N.; Jeong, M.-J.; Park, J.; Shin, K.S.; Kim, T.; Kim, N. A Rectangular Notch-Band UWB Antenna with Controllable Notched Bandwidth and Centre Frequency. *Sensors* **2020**, *20*, 777. [CrossRef]
22. Al-Bawri, S.S.; Hwang Goh, H.; Islam, M.S.; Wong, H.Y.; Jamlos, M.F.; Narbudowicz, A.; Jusoh, M.; Sabapathy, T.; Khan, R.; Islam, M.T. Compact ultra-wideband monopole antenna loaded with metamaterial. *Sensors* **2020**, *20*, 796. [CrossRef] [PubMed]
23. Lakrit, S.; Das, S.; Madhav, B.T.P.; Babu, K.V. An octagonal star shaped flexible UWB antenna with band-notched characteristics for WLAN applications. *J. Instrum.* **2020**, *15*, P02021. [CrossRef]

24. Jameel, M.S.; Mezaal, Y.S.; Atilla, D.C. Miniaturized coplanar waveguide-fed UWB Antenna for wireless applications. *Symmetry* **2023**, *15*, 633. [CrossRef]
25. Kumar, P.; Ali, T.; MM, M.P. Characteristic mode analysis-based compact dual band-notched UWB MIMO antenna loaded with neutralization Line. *Micromachines* **2022**, *13*, 1599. [CrossRef] [PubMed]
26. Jaglan, N.; Gupta, S.D.; Kanaujia, B.K.; Srivastava, S. Band notched UWB circular monopole antenna with inductance enhanced modified mushroom EBG structures. *Wirel. Netw.* **2018**, *24*, 383–393. [CrossRef]
27. Xu, H.; Xu, K.D.; Nie, W.; Liu, Y.H. A coplanar waveguide fed UWB antenna using embedded E-shaped structure with WLAN band-rejection. *Frequenz* **2018**, *72*, 325–332. [CrossRef]
28. Aitbar, I.; Shoaib, N.; Alomainy, A.; Quddious, A.; Nikolaou, S.; Imran, M.A.; Abbasi, Q.H. AMC integrated multilayer wearable antenna for multiband WBAN applications. *Comput. Mater. Contin.* **2022**, *71*, 3227–3241. [CrossRef]
29. Kumar, O.P.; Kumar, P.; Ali, T. A compact dual-band notched UWB antenna for wireless applications. *Micromachines* **2021**, *13*, 12. [CrossRef]
30. Awan, W.A.; Hussain, N.; Kim, S.; Kim, N. A frequency-reconfigurable filtenna for GSM, 4G-LTE, ISM, and 5G Sub-6 GHz band applications. *Sensors* **2022**, *22*, 5558. [CrossRef]
31. Iqbal, A.; Smida, A.; Mallat, N.K.; Islam, M.T.; Kim, S. A compact UWB antenna with independently controllable notch bands. *Sensors* **2019**, *19*, 1411. [CrossRef]
32. Mukherjee, K.; Mukhopadhyay, S.; Roy, S. Design of a wideband Y-shaped antenna for the application in IoT and 5G communication. *Int. J. Commun. Syst.* **2022**, *35*, 5021. [CrossRef]
33. Jeong, M.J.; Hussain, N.; Bong, H.U.; Park, J.W.; Shin, K.S.; Lee, S.W.; Rhee, S.Y.; Kim, N. Ultrawideband microstrip patch antenna with quadruple band notch characteristic using negative permittivity unit cells. *Microw. Opt. Technol. Lett.* **2020**, *62*, 816–824. [CrossRef]
34. Mekki, K.; Necibi, O.; Lakhdhar, S.; Gharsallah, A. A UHF/UWB monopole antenna design process integrated in an RFID reader board. *J. Electromag Eng. Sci.* **2022**, *22*, 479–487. [CrossRef]

applied sciences

MDPI

Article

Design of a Wide-Bandwidth, High-Gain and Easy-to-Manufacture 2.4 GHz Floating Patch Antenna Fed with the Through-Wire Technique

Marcos D. Fernandez [1,*], Darío Herraiz [1], David Herraiz [2], Akram Alomainy [3] and Angel Belenguer [1]

1 Departamento de Ingeniería Eléctrica, Electrónica, Automática y Comunicaciones, Universidad de Castilla-La Mancha, Escuela Politécnica de Cuenca, Campus Universitario, 16071 Cuenca, Spain;
2 Instituto de Telecomunicaciones y Aplicaciones Multimedia, Universitat Politècnica de València, Camino de Vera, s/n, 46022 Valencia, Spain
3 School of Electronic Engineering and Computer Science, Queen Mary University of London, London E1 4NS, UK
* Correspondence: marcos.fernandez@uclm.es

Abstract: This paper presents a feasibility study for designing a floating patch antenna structure fed with a probe from a microstrip. The main premise is to eliminate the dielectric in the patch design, which is equivalent to having an air dielectric and leads to the necessity of proper support to fasten the patch in the air. The novelty of this paper is that this new device, apart from being fed with the though-wire technique directly from the microstrip line, has to be, by design, robust and easy to manufacture, and, at the same time, it has to present, simultaneously, good values in all of the performance indexes. A prototype has been designed, manufactured, and measured with good performance results: a bandwidth higher than 10% around 2.4 GHz, a radiation efficiency higher than 96%, a 9.63 dBi gain, and a wide beamwidth. The main advantages of this prototype, together with its good performance indexes, include its low fabrication cost, low losses, light weight, robustness, high integration capability, the complete removal of the dielectric material, and the use of a single post for feeding the patch while simultaneously fixing its floating position.

Keywords: floating patch; antenna; air substrate; microstrip; through-wire

Citation: Fernandez, M.D.; Herraiz, D.; Herraiz, D.; Alomainy, A.; Belenguer, A. Design of a Wide-Bandwidth, High-Gain and Easy-to-Manufacture 2.4 GHz Floating Patch Antenna Fed with the Through-Wire Technique. *Appl. Sci.* **2022**, *12*, 12925. https://doi.org/10.3390/app122412925

Academic Editor: Ernesto Limiti

Received: 8 November 2022
Accepted: 14 December 2022
Published: 16 December 2022

1. Introduction

The quick development of wireless communication systems for the general public implies the improvement of the radiation elements, among which the microstrip patch antenna has attracted much attention because of its main advantages: its lightweight, low-profile nature; its low-cost, easy fabrication process; etc. Nevertheless, it also presents some drawbacks, such as its narrow bandwidth; low gain; poor polarization purity; limited power capacity; etc. [1].

Essentially, a microstrip patch antenna is made up of a patch on one side of a dielectric substrate and a ground plane on the other side of the substrate; the feeding is made through a microstrip line, whether directly with insets, with an aperture-coupled probe, or with a coaxial probe. The patch may be of any geometry (it is commonly rectangular and circular) [2].

In fact, it is usually recognized that it is quite challenging to design an antenna that combines a high-bandwidth, high-gain, and wide beamwidth with the properties of low cost, lightweight, and ease of manufacture; some attempts to achieve this are reported in the scientific literature by using substrate-integrated waveguides, multipatch designs, multiple slots feeding the patch, a multi-layered substrate, variations in the feeding structure, the use of lumped elements, a combination of dielectrics (as in dielectric resonator antennas), different patch shapes, the inclusion of parasitic elements over the patch, etc. [3–7].

In [8], the authors state that the optimal selection of the dielectric material and its thickness has little influence on the resonance frequency, but it is of paramount importance to achieve maximum radiation efficiency and bandwidth. The dielectric constant also affects the size of the patch: the higher the dielectric constant, the tinier the patch will be. This fact is used in [9], in which an antenna array at 2.45 GHz with a high relative permittivity substrate has been used to minimize the size of a human wearable device. Nevertheless, theoretically, the smaller the dielectric constant, the better the antenna performance will be [10]. In fact, in [11–14], researchers proved theoretically and experimentally that a floating patch with an air dielectric improves the performance of the bandwidth and gain; however, this is at the cost of a small increase in size. This increment can be minimized by shorting posts and slots at the cost of increasing the manufacturing complexity [15].

Some examples of dielectric removal by using air as the separation between the floating patch and the feeding network have been reported in the literature; for instance, in [16], a suspended patch based on a microstrip coaxial probe feed technique was presented with an excellent radiation efficiency higher than 90%; however, this solution presents a poor fractional bandwidth of lower than 5%. On the other hand, in [17], a transparent water patch fed by an L-shaped probe was presented with a fractional bandwidth higher than 30% but a poor radiation efficiency of lower than 80% and higher manufacturing complexity. Another example, in [8], a floating patch fed by a microstrip and inserts was presented with a radiation efficiency of higher than 90% but a poor fractional bandwidth of less than 6%. As another example, in [18], a suspended conformal patch antenna was proposed, and the authors compared it to the non-conformal version, both of which include a non-floating patch and an air cavity backed between the patch substrate and the reflection ground. The best results give good values for the gain but a poor bandwidth, apart from the necessity of a supporting frame to make the air cavity. Furthermore, in [19,20], examples of slot-strip-foam-inverted patch (RSSFIP) antennas and arrays are proposed, in which the single element antenna achieves an excellent gain value, but the device requires an H-shaped slot, two layers of substrate, and an additional metallic reflector layer, all of which are separated by two air layers that, in fact, increase the complexity of the manufacturing process. On the whole, this reveals that, for a single radiating element patch antenna, getting a bandwidth of higher than 10% with a radiation efficiency of more than 90% is very complex, and it depends on the feeding technique and the floating patch construction design.

All of the former examples make use of different techniques that include air in the improvement of the performance indexes, whether as a cavity-backed [18] or additional layer, such as in the air-gap technique [13], the suspended patch technique [21], or the inverted patch technique [19]. However, in none of these is the patch really isolated as a completely dielectricless floating element only surrounded by air, which is the new approach used in this novel proposal. In order to solve the above challenges, this work deals with the aim of designing a microstrip patch antenna with an air substrate in the simplest possible way (complex antenna designs are not of interest to the rapidly growing wireless industries [8]). It needs to be able to cope with a high gain, radiation efficiency, low loss, and wide bandwidth, which are the common performance requirements of current radiating elements, while maintaining its compact size, light weight, ease of manufacture, and easy integration with other electronic devices [22]. Therefore, in order to cope with these requirements, this work proposes a real dielectricless floating patch in which the dielectric is removed by using air ($\epsilon_{r_{air}} = 1$) as the separation between the floating patch and the feeding. Next, the patch shape is chosen to be the simplest for manufacturing, i.e., a rectangle (as this shape minimizes cuts in the base material and can be completed with maximum accuracy). Lastly, and most importantly, the patch feeding is adjusted through a pin rod (probe) to the microstrip feeding line, taking advantage of the through-wire technique equivalently used in [23,24] for feeding an empty substrate-integrated waveguide. This new feeding technique is used for the first time to feed a patch antenna; it has the same advantages as the coaxial feed, together with low losses, a perfect adaptation to the microstrip line, and a straight and very simple way of connecting to the patch with just a direct insertion point to be tin-soldered (as this is just a monoconductor pin rod). This pin rod will serve, at the same time, as the fastener of the patch to fix its floating position, which therefore is easier with regard to the manufacturing process. The proper combination of these pieces allows for the achievement of an extremely good-performance patch antenna fulfilling all of the desired requirements, and, under a new design approach characterized by a completely dielectricless patch antenna, uses the new through-wire technique to feed and hold the real floating patch and make it through a simple structure with a very quick and easy manufacturing procedure.

2. Design Goals and Procedure

The proposed antenna radiates at 2.4 GHz, with a desired 10% minimum bandwidth and 95% minimum radiation efficiency, through a floating metallic thin patch with h_{air} mm of an air dielectric ($\epsilon_{r_{air}} = 1$) between the patch and the ground plane. The patch is implemented with a fully metalized layer of a substrate of h_{patch} mm total thickness (considering the 17 μm of cladding and 9 μm of copper electrodeposition to completely metalize the whole patch); in fact, the patch can also be completed with just a copper sheet of h_{patch} thickness and sufficient stiffness. For the feeding section, a microstrip line of a w_{ms} mm width is used; it is set in another layer of substrate (Rogers-4003C, $\epsilon_{r_{subst}} = 3.55$, $\tan\delta = 0.0027$, and 17 μm of cladding for each side) of an h_{ms} mm thickness. A proper fastener to hold and stabilize the floating patch is needed, which is the feeding pin rod that injects the power into the patch from the microstrip line; the design and adjustment of this element are critical for achieving a good impedance matching and a successful device with high values of efficiency. The well-known design equations for a microstrip patch included in Chapter 14 of [10] and summarized in [25] are used to calculate the initial values of the patch size; the used equations give only an initial approximation of the dimensions, so the device is modeled and numerically evaluated through optimization algorithms using CST Studio Suite to obtain the final dimensions and the adjustment of the pin rod's position in the patch.

The appearance and design parameters of the prototype are illustrated in Figure 1. The three parts composing the prototype are shown: the lower layer with a microstrip line whose ground plane is also used as the ground plane for the patch; the metallic floating patch optimized to radiate at 2.4 GHz; and the pin rod holding the patch and feeding it from the microstrip, whose insertion position and isolation from the ground plane are optimized to maximize the impedance matching.

The patch is just a completely metalized piece of Rogers-4003C of a thickness of 0.813 mm, giving a total thickness of $h_p = 0.866$ mm; its shape is a rectangle of the size $L_p \times W_p$ at the distance h_{air} over the ground of the microstrip layer. This distance is chosen to be $h_{air} = 8.13$ mm to ease the manufacture (it is the piling of ten substrates of a height of 0.813 mm). For a chosen radiation frequency of $f_r = 2.4$ GHz and the air as the patch dielectric ($\epsilon_{r_{eff}} = \epsilon_{r_{air}} = 1$), the design equations for the patch are simplified as follows:

$$W_p = \frac{c}{2f_r} \tag{1}$$

$$\Delta L = 0.722 h_{air} \frac{0.264 + \frac{W_p}{h_{air}}}{0.8 + \frac{W_p}{h_{air}}} \tag{2}$$

$$L_p = \frac{c}{2f_r} - 2\Delta L \tag{3}$$

where $W_p = 62.457$ mm is the size of the radiating edges (y axis), $L_p = 51.459$ mm is the size of the non-radiating edges (x axis), $\epsilon_{r_{eff}}$ is the effective dielectric constant (that matches $\epsilon_{r_{air}} = 1$ in this case), and $\Delta L = 5.499$ mm is the correction in size due to the fringing effects on the patch. The size of the patch reference plane ($L_s \times W_s$) must be greater than the patch; it is recommended to be at least approximately the size of the patch incremented six times by the value of the dielectric ($L_s = L_p + 6h_{air}$, $W_s = W_p + 6h_{air}$).

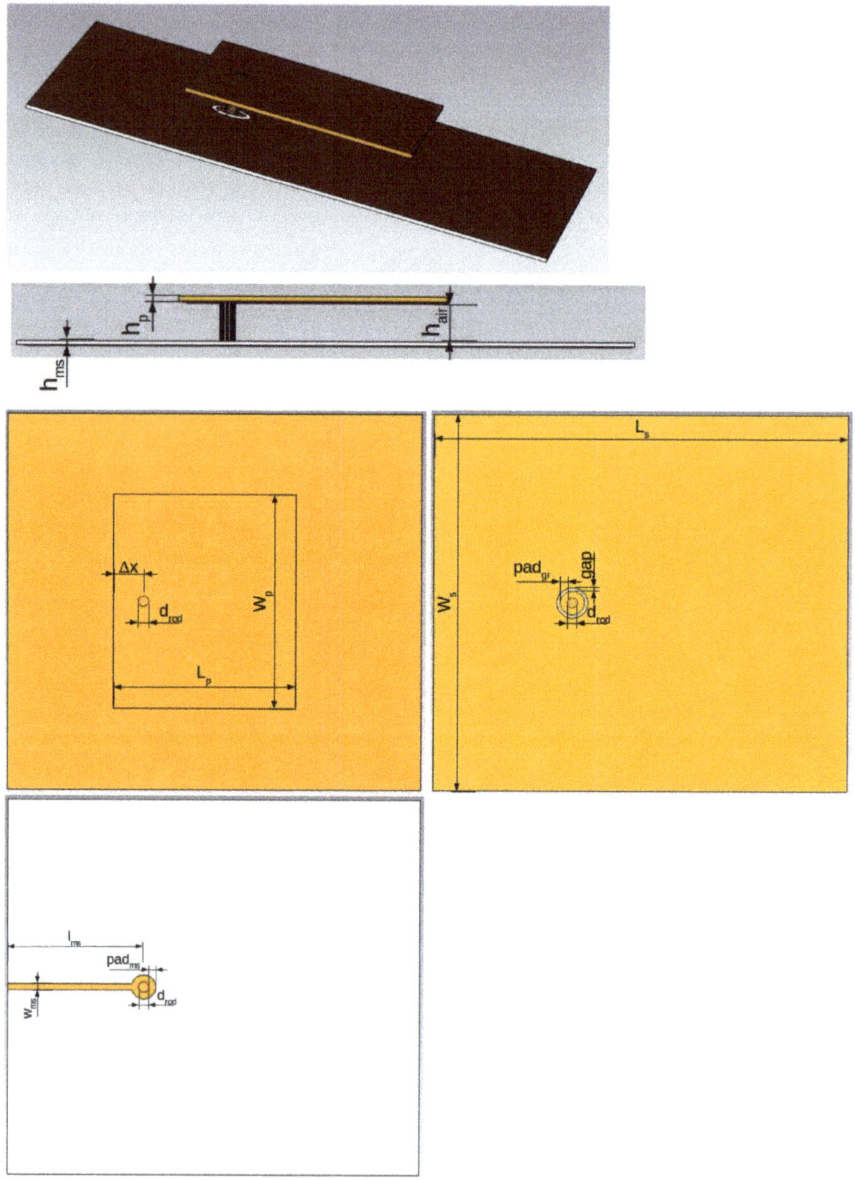

Figure 1. Appearance and design parameters of the floating patch prototype. From top-left to bottom-right: 3D view from the top, lateral view of the device, top view of the device, ground side of the microstrip layer, and line view of the microstrip layer. Main dimensions: $L_s = 120$ mm, $W_s = 110$ mm, $L_p = 52.586$ mm, $W_p = 62.457$ mm, $d_{rod} = 3$ mm, and $h_{air} = 8.13$ mm. Color legend: white is the substrate, dark yellow/brown is copper, and background gray is air.

The microstrip layer is integrated into an $h_{ms} = 0.813$ mm thick substrate of the size $L_s = 120$ and $W_s = 110$ mm with a 50 Ω impedance microstrip line on the bottom side of the length $l_{ms} = 40$ mm and width $w_{ms} = 1.784$ mm. The top side of this layer is the common ground plane for the microstrip and patch.

The feeding of the patch is made with a coaxial probe that has been implemented as a pin rod connecting the microstrip line with the patch through the ground plane by using the through-wire technique stated in [23]. With the aim of using the pin rod as the holding post of the patch, and considering the size of the patch, a 3 mm diameter metallic rod has been selected to ensure the robustness and stability of the device. To make the manufacturing process easier, the values of the pads and gaps necessary for the through-wire technique have been forced to the values of 1 mm for the gap in the ground plane and 2 mm for the pads around the metallic rod. As the rod must trespass the microstrip layer, a through-hole must be made; that hole needs to be metalized (with 9 μm of copper electrodeposition) to achieve good connectivity with the rod, so that procedure needs to be completed before printing the microstrip line.

The insertion position of the pin rod into the patch is critical. According to the considerations stated in Chapter 21.6 of [26], the insertion position for a linearly polarized antenna should be at the middle of one of the radiating edges and separated a distance Δx from the edge toward the center of the patch. It is well-known that the impedance strongly varies along the antenna, and, furthermore, the impedance and bandwidth are dependent on the ratio W_p/L_p because it changes the quantity of energy stored in the cavity formed by the patch and the ground plane. Consequently, it is stated in [26] that the precise insertion position is very sensitive to L_p and Δx and only initial approximations can be given that must later be adjusted empirically through a simulator. Taking the former into account and considering that the air is the dielectric of the patch, the initial position of the insertion pin rod is estimated by $\Delta x \approx 0.25L_p$.

This device has been simulated (including losses), and the values of Δx and ΔL (and therefore L_p) have been optimized for adjustment of the insertion position of the pin rod that maximizes the radiation at 2.4 GHz. The optimum point indicates that the antenna has a bandwidth of 10.46% around 2.4 GHz (according to the band with $S_{11} < -10$ dB), with a directivity of 9.95 dBi and a total efficiency of 99.78%; those values are quite promising, as they reflect the high-performance indexes of the bandwidth, efficiency, and gain at the same time for just a single radiating element. The final dimensions of the optimized device are those in Table 1.

Table 1. Optimum dimensions (in mm) for the floating patch device.

l_{ms}	w_{ms}	h_{ms}	d_{rod}	pad_{ms}
40	1.784	0.813	3	2
L_s	W_s	pad_{gr}	gap	h_{air}
120	110	2	1	8.13
ΔL	L_p	W_p	Δx	h_p
4.936	52.586	62.457	8.862	0.866

3. Results and Discussion

The designed device has been manufactured, and the resulting prototype is shown in Figure 2. Just two pieces of the substrate are needed: one for the microstrip layer and another for the patch, apart from a short metallic rod to fasten both layers. The manufacturing process is quite simple and can be completed with standard PCB operations, such as drilling, cutting, milling, plating, and soldering. The hole for the rod is drilled into both substrate pieces. Then, the microstrip layer is plated to metalize the hole. Afterward, the microstrip line is printed, and the final layer is cut. On the other hand, after drilling the patch layer, it is directly cut and completely plated. Next, the rod is cut to the desired length, and it is tin-soldered to the patch. Finally, the rod has to be carefully soldered to the microstrip layer.

Figure 2. Manufactured prototype; top and frontal views with a reference scale.

The prototype has been measured with a vector network analyzer (Agilent N5230C) to characterize the S_{11} parameter. The result is shown in Figure 3, which compares the measurement to the simulation. It reflects practically a total match between the simulated device and the manufactured prototype; the measured return loss is 30 dB at 2.4 GHz with a measured bandwidth of 10.41% around the resonance frequency. This fulfills one of the goals of achieving a high bandwidth for the patch antenna by using the air as a dielectric and a through-wire feed.

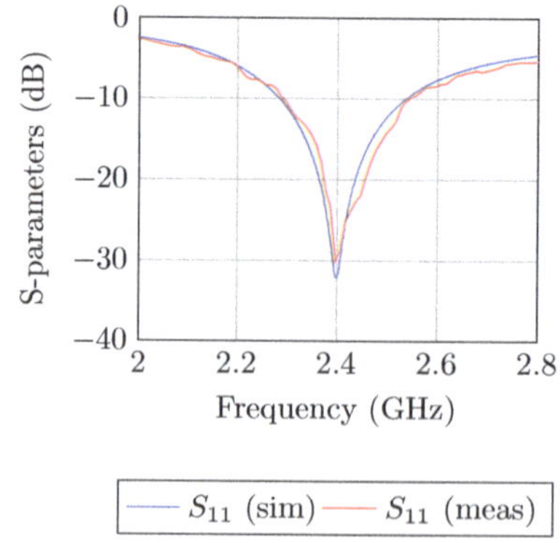

Figure 3. Comparison of the measured and simulated parameter S_{11}.

The radiation of the prototype patch antenna has been also measured in an anechoic chamber with the technique of the reference antenna (QRH500, in this case), as can be seen in Figure 4. With the data acquired in this measurement procedure, the antenna can be completely characterized.

Figure 4. Measurement of the prototype in an anechoic chamber.

The measured radiation pattern is shown in Figure 5, whose XZ ($\phi = 0°$), YZ ($\phi = 90°$), and XY ($\theta = 90°$) plane cuts are included in Figure 6. The obtained patterns are typical of a patch antenna. The measured directivity at the resonance frequency is 9.94 dBi, with a measured gain of 9.63 dBi, which gives a radiation efficiency of 96.41% and a total efficiency of 93.08%. The maximum direction of the radiation is $\theta = 2°$ and $\phi = 224°$, whereas the -3 dB beamwidth is 52° for plane YZ and 60° for the XZ. Moreover, a good cross-polar rejection is obtained, with 67 dB for the polarization ellipse axial ratio. These results fit almost perfectly with the simulations and reveal, as intended, the simultaneous high-performance indexes of the return loss, bandwidth, beamwidth, gain, and efficiency. c

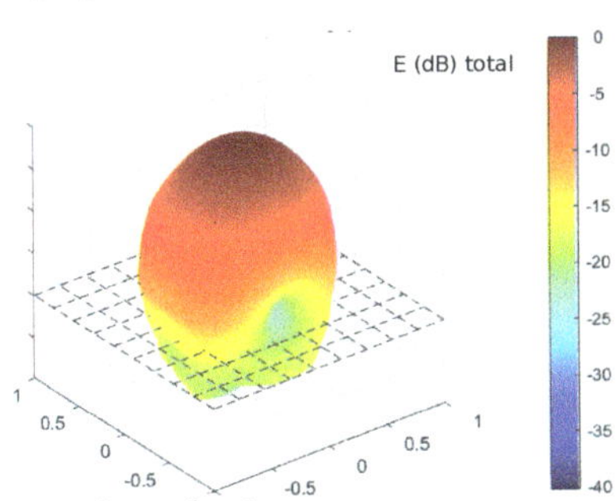

Figure 5. 3D representation of measured electric field radiation.

Despite the quite easy manufacturing process for this antenna, some inherent fabrication errors may occur. A yield analysis has been performed to check the endurance and repeatability of the proposed prototype and its manufacturing. It has been considered that errors may appear in the most critical operations: the milling pads and gap for the rod as well as the drilling position for the rod in the patch and the cutting process for the size of the patch. The selected criteria to mark a device as acceptable are that $S_{11} < -10$ dB at a minimum bandwidth of 10% around the resonance frequency. Taking into account that the tolerance of the laser cuts used is 2 microns, the analysis indicates that, even for high tolerances of 25 microns, 100% of the devices would be acceptable. If the tolerance were extremely high, for instance, 50 microns, 99% of the prototypes would be acceptable; even for the unlikely case of an excessively high tolerance of 100 microns, 96% of the prototypes could be still acceptable. This reveals the accuracy of the design and the high tolerance for manufacturing errors of this prototype.

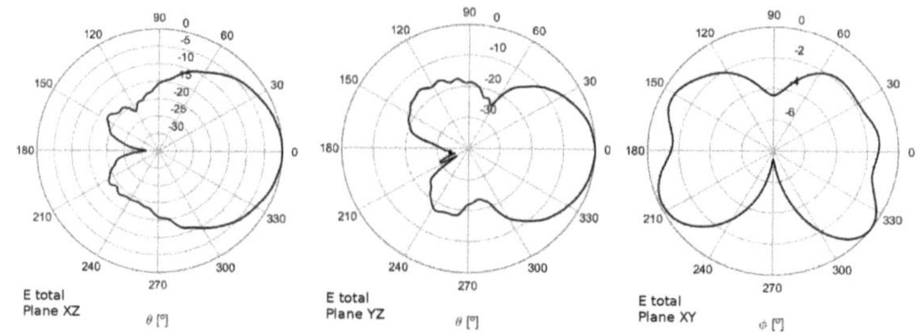

Figure 6. Polar plot cuts of measured electric field radiation. From left to right: plane XZ, plane YZ, and plane XY.

The results above prove the feasibility of this prototype and validate the simplicity and endurance of its fabrication procedure. To contextualize this new patch antenna device, the performance indexes of different air substrate microstrip patch antennas with just one radiating element are compared in Table 2 regarding the resonance frequency, the return loss, the fractional bandwidth, the gain, the efficiency, and the feeding technique; qualitative aspects about the manufacturing complexity and other details of the prototypes are also included. The data in the table show that every published antenna can stand out in one or two performance indexes, and it also clearly reveals that this new antenna is the only one that is simultaneously completely dielectricless, able to achieve good values at the same time in all of the performance indexes, very easy to manufacture, and uses just one holder for both feeding the patch and fixing its position. The same happens if this antenna is compared with other patch antennas using different substrates, such as those in the compilation included in [27].

Table 2. Comparison of measured performance parameters of air substrate patch antennas.

Ref.	f_r (GHz)	R_L (dB)	BW (%)	G (dBi)	η (%)	Feeding	Manufacturing Complexity	Comments
[16]	0.445	32.00	4.50	8.55	95.00	mstrip-coax	easy	only simulation, aluminium patch, eight holders
[17]	1.9	—	34.90	6.60	75.00	coax-L_probe	very difficult	water-based, transparent patch
[21]	2	24.00	7.00	6.60	—	coax-mstrip	very difficult	complex structure, dual polarization
[11]	2.44	30.00	4.94	9.78	99.00	coax	easy	additional metallic holding post, η simulated
[14]	2.45	23.00	5.00	7.91	—	mstrip	easy	air-gap, non-dielectricless
[15]	2.45	26.50	5.00	3.90	89.00	coax	medium	shorting posts, slots
[18]	2.45	20.00	3.00	8.51	—	mstrip-insets	medium	air cavity backed
[28]	2.44	21.00	—	8.40	88.00	mstrip-coax	difficult	dual band, semi-circular slot, four holders
This	**2.4**	**30.00**	**10.41**	**9.63**	**96.41**	**mstrip-through_wire**	**very easy**	**single holder, completely dielectricless**

4. Conclusions

In this paper, the design, fabrication, and measurement of a floating patch antenna are proposed. The design procedure is eased by selecting air as the dielectric for the patch, as well as a simple patch shape; the most critical aspect is the positioning of the feeding pin rod that connects the microstrip line to the patch so that there is a good impedance matching able to maximize the radiation at the resonance frequency. The manufacturing processes are standard PCB operations, and the whole fabrication procedure is very easy and highly tolerant of manufacturing errors. This antenna is able to radiate at 2.4 GHz with a 10.41% fractional bandwidth, 9.63 dBi gain, 96.41% radiation efficiency, and total efficiency of 93.08%, together with a wide beamwidth and good polarization purity. Thanks to the microstrip and through-wire feeding system, this antenna presents a high integration capability with other planar circuits; at the same time, this device is low-cost, lightweight, robust, and easy to manufacture. The main novelties of this design are its ease of design and manufacture, the total absence of a dielectric, the use of the through-wire technique to feed the patch and fasten the floating patch, the extreme robustness and error tolerance of the prototype, and the simultaneous achievement of good values in all of the performance indexes. The proposed antenna can be profitably used for wireless applications in the 2.4 GHz band.

Author Contributions: M.D.F., A.A. and A.B. conceptualized the design; M.D.F. and A.B. designed, simulated, and optimized the device; D.H. (Darío Herraiz) and D.H. (David Herraiz) manufactured the device and characterized it in the laboratory; M.D.F. wrote the manuscript; all authors revised the manuscript; A.B. and A.A. supervised the whole work. All authors have read and agreed to the published version of the manuscript.

Funding: This work was supported by the Ministerio de Ciencia e Innovación of the Spanish Government through the Subproject C44 of the Coordinated Research and Development Project PID2019-103982RB under Grant MCIN/AEI/10.13039/501100011033.

Institutional Review Board Statement: Not applicable.

Informed Consent Statement: Not applicable.

Data Availability Statement: All data are available upon reasonable request.

Conflicts of Interest: The authors declare no conflict of interest.

References

1. Tamim, A.M.; Sunbeam Islam, S.; Rahman Chowdhury, M.A.; Hossain, M.I. Design of a High Gain Microstrip Patch Antenna for X-band Satellite Applications. In Proceedings of the 2018 International Conference on Computer, Communication, Chemical, Material and Electronic Engineering (IC4ME2), Rajshahi, Bangladesh, 8–9 February 2018; pp. 1–5. [CrossRef]
2. Gupta, M.; Sachdeva, S.; Swamy, N.K.; Singh, I. Rectangular Microstrip Patch Antenna Using Air as Substrate for S-Band Communication. *J. Electromagn. Anal. Appl.* **2014**, *6*, 38–41. [CrossRef]
3. Mohammed, A.S.; Kamal, S.; Ain, M.F.B.; Najmi, F.; Ahmad, Z.A.; Zahar, Z.; Hussin, R.; Zubir, I.A.; Rahman, M.A. Improving the Gain Performance of Air Substrate Patch Antenna Array Using the Effect of Conductive Material Thickness Study for 5G Applications. *J. Phys. Conf. Ser.* **2020**, *1529*, 052020. [CrossRef]
4. Gemeda, M.; Fante, K.; Goshu, H.; Goshu, A. Design and Analysis of a 28 GHz Microstrip Patch Antenna for 5G Communication Systems. *Int. Res. J. Eng. Technol.* **2021**, *8*, 881–886.
5. Moayyed, F.; Oskouei, H.; Mohammadi Shirkolaei, M. High Gain and Wideband Multi-Stack Multilayer Anisotropic Dielectric Antenna. *Prog. Electromagn. Res. Lett.* **2021**, *99*, 103–109. [CrossRef]
6. Farad, M.M.; Dixit, M. Bandwidth Enhancement of Microstrip patch Antenna Using Suspended Techniques for Wireless Applications. *Int. J. Eng. Res. Technol.* **2014**, *121*, 103–120.
7. Anim, K.; Danuor, P.; Park, S.; Jung, Y. High-Efficiency Broadband Planar Array Antenna with Suspended Microstrip Slab for X-Band SAR Onboard Small Satellites. *Sensors* **2022**, *22*, 252. [CrossRef]
8. Mohammed, A.S.; Kamal, S.; Bin Ain, M.F.; Hussin, R.; Najmi, F.; @Suandi, S.A.S.; Arifin Ahmad, Z.; Ullah, U.; Bin Mohamed Omar, M.F.; Othman, M. Mathematical model on the effects of conductor thickness on the centre frequency at 28 GHz for the performance of microstrip patch antenna using air substrate for 5G application. *Alex. Eng. J.* **2021**, *60*, 5265–5273. [CrossRef]
9. Shirkolaei, M.M.; Oskouei, H.R.D.; Abbasi, M. Design of 1*4 Microstrip Antenna Array on the Human Thigh with Gain Enhancement. *IETE J. Res.* **2021**.

10. Balanis, C.A. *Antenna Theory: Analysis and Design*; John Wiley: Hoboken, NJ, USA, 2016.
11. Belamgi, S.B.; Ray, S.; Das, P. Suspended planar patch antenna for wireless energy transfer at 2.45 GHz. In Proceedings of the International Conference on Electronics, Communication and Instrumentation (ICECI), Kolkata, India, 16–17 January 2014; pp. 1–4. [CrossRef]
12. Abdullah, M.S.; Mardeni, R. A Novel of Suspended Plate Compact Antenna Design for 2.4 GHz Applications. *Mod. Appl. Sci.* **2014**, *8*, 258. [CrossRef]
13. Verma, R.; Saxena, N.; Pourush, P. Effect of air-gap technique in bandwidth of microstrip patch array antenna. *Int. J. Res. Publ. Eng. Technol.* **2017**, *3*, 165–168.
14. Al Kharusi, K.W.S.; Ramli, N.; Khan, S.; Ali, M.T.; Halim, M.A. Gain Enhancement of Rectangular Microstrip Patch Antenna using Air Gap at 2.4 GHz. *Int. J. Nano Mater.* **2020**, *13*, 211–224.
15. Belamgi, S.B.; Ray, S. A compact suspended planar patch antenna for microwave imaging sensor array. In Proceedings of the 2015 Third International Conference on Computer, Communication, Control and Information Technology (C3IT), Hooghly, India, 7–8 February 2015; pp. 1–4. [CrossRef]
16. Elluru, J.; Sourirajan, V. Design of UHF-band Microstrip patch Antenna for Wind Profiler Radar. *Int. J. Adv. Eng. Res. Dev.t* **2017**, *4*, 567–572.
17. Sun, J.; Luk, K.M. A Wideband and Optically Transparent Water Patch Antenna with Broadside Radiation Pattern. *IEEE Antennas Wirel. Propag. Lett.* **2020**, *19*, 341–345. [CrossRef]
18. Meerabeab, S.; Jantarachote, V.; Wounchoum, P. Design and Parametric Study of a Suspended Conformal Patch Antenna. In Proceedings of the 2022 19th International Conference on Electrical Engineering/Electronics, Computer, Telecommunications and Information Technology (ECTI-CON), Prachuap Khiri Khan, Thailand, 24–27 May 2022; pp. 1–4. [CrossRef]
19. Mohammadi Shirkolaei, M. Wideband linear microstrip array antenna with high efficiency and low side lobe level. *Int. J. Microw. Comput. Aided Eng.* **2020**, *30*, e22412. [CrossRef]
20. Mohammadi Shirkolaei, M. High Efficiency X-Band Series-Fed Microstrip Array Antenna. *Prog. Electromagn. Res. C* **2020**, *105*, 35–45. [CrossRef]
21. Lin, Y.M.; Yang, J.S.; Row, J.S. A design for suspended patch antenna with switchable polarization. *Microw. Opt. Technol. Lett.* **2016**, *58*, 1333–1337. [CrossRef]
22. Mittra, R. Some Challenges in Millimeter Wave Antenna Designs for 5G. In Proceedings of the 2018 International Symposium on Antennas and Propagation (ISAP), Busan, Republic of Korea, 23–26 October 2018; pp. 1–2.
23. Belenguer, A.; Ballesteros, J.; Fernandez, M.; Esteban, H.; Boria, V. Versatile, error-tolerant and easy to manufacture through-wire microstrip-to-ESIW transition. *IEEE Trans. Microw. Theory Tech.* **2020**, *68*, 2243–2250. [CrossRef]
24. Fernandez, M.D.; Ballesteros, J.A.; Belenguer, A. Highly Compact Through-Wire Microstrip to Empty Substrate Integrated Coaxial Line Transition. *Appl. Sci.* **2021**, *11*, 6885. [CrossRef]
25. Przesmycki, R.; Bugaj, M.; Nowosielski, L. Broadband Microstrip Antenna for 5G Wireless Systems Operating at 28 GHz. *Electronics* **2020**, *10*, 1. [CrossRef]
26. Lee, T.H. *Planar Microwave Engineering: A Practical Guide to Theory, Measurements and Circuits*; Cambridge University Press: Cambridge, UK, 2004.
27. Christina, G. A Review on Microstrip Patch Antenna Performance Improvement Techniques on Various Applications. *J. Trends Comput. Sci. Smart Technol.* **2021**, *3*, 175–189. [CrossRef]
28. Ou, J.H.; Andrenko, A.S.; Li, Y.; Zhang, Q.; Tan, H.Z. High-Efficiency and Wide-Frequency-Ratio Dual-Band Slot Patch Antenna Utilizing the Perturbed TM02 Modes. *IEEE Antennas Wirel. Propag. Lett.* **2018**, *17*, 579–582. [CrossRef]

applied sciences

MDPI

Article

Simulation-Based Selection of Transmitting Antenna Type for Enhanced Loran System in Selected Location

Zhaopeng Hu [1,*], Ming Dong [2,*], Shifeng Li [1] and Chanzhong Yang [1]

[1] National Time Service Centre, Chinese Academy of Sciences, Xi'an 710600, China; lishifeng@ntsc.ac.cn (S.L.); yangcz@ntsc.ac.cn (C.Y.)
[2] Beijing Institute of Tracking and Telecommunications Technology, Beijing 100094, China
* Correspondence: huzhaopeng@ntsc.ac.cn (Z.H.); dongmingdmdm1985@126.com (M.D.)

Abstract: To improve the coverage and timing capability of enhanced Loran signals, three enhanced Loran transmitters are planned to be built in Western China. The most appropriate antenna configuration can be determined by comparing domestic and foreign mainstream small and radio antennas. By analyzing and comparing the electrical and structural parameters and signal propagation curves of the transmitting antenna, it can be concluded that the single tower umbrella antenna provides the best performance in all evaluation indexes, and the enhanced Loran signal can be transmitted to areas 1000 km away through the single tower umbrella antenna so that the enhanced Loran signal covers most areas of Western China. Therefore, it should be widely used in the construction of enhanced Loran transmitters in the future.

Keywords: enhanced loran signal; transmitting antenna; propagation curve

Citation: Hu, Z.; Dong, M.; Li, S.; Yang, C. Simulation-Based Selection of Transmitting Antenna Type for Enhanced Loran System in Selected Location. *Appl. Sci.* **2022**, *12*, 6031. https://doi.org/10.3390/app12126031

Academic Editor: Naser Ojaroudi Parchin

Received: 19 April 2022
Accepted: 9 June 2022
Published: 14 June 2022

1. Introduction

The enhanced Loran system evolved from the Loran-C (long-range navigation) system. It is an internationally standardized positioning, navigation and timing (pnt) service system. It is the latest development achievement of low-frequency long-range radio navigation timing technology. It meets the needs of PNT services in aviation non-precision approach, port approach, land vehicle navigation and location service in terms of PNT service accuracy, reliability, effectiveness, perfection and continuity. Additionally, it can provide high-precision time and frequency signals in the order of hundreds of nanoseconds. The eLoran transmitting antenna is an important part of the system; its electrical performance and coverage ability are very important [1–3].

Loran transmitting antennas are a mature technology has that has benefitted from many years of research and development. Typically, they comprise single-tower umbrella antennas or four-tower inverted cone antennas [4,5].

China plans to build three enhanced Loran launchers in the western region. Its purpose is to cover most parts of China with Loran time service signal, which will be used for time service and navigation in the future. At present, the performance of the Loran-C system in all aspects of operation is relatively backward. Therefore, the three new transmitting stations will choose the Loran transmitting antenna with higher performance. In particular, the Tibet Naqu transmitting station, one of the three largest transmitting stations, will realize the construction of Loran transmitting antennas for the first time in regions above 4500 m above sea level in the world. Referring to the past design and construction experience of Loran transmitting antennas, it is difficult to meet the current needs. Therefore, the design and selection of the transmitting antenna will face new challenges. The antenna not only needs to cope with the complex geographical characteristics of Tibet, but also needs to meet the requirements of the transmitter and improve the signal radiation capacity as much as possible, which will put forward higher requirements for the structural and electrical performance design of the transmitting antenna. Its main purpose is to carry the signal transmitted by a 2500 kW high-power transmitter and radiate it efficiently. The radiation efficiency is more than 70%. It is planned to cover most of the land area in Western China. The main performance requirements are as follows: (1) the working center frequency is 100 kHz; (2) The bandwidth exceeds 3.5 kHz; (3) The effective height exceeds 180 m; (4) The characteristic impedance is 200–300 Ω [6,7].

2. International Typical Transmitting Antenna Reference

2.1. China

The BPL long-wave timing station is located in Pucheng County, Weinan City, Shaanxi Province. It is the first large-scale radio station in China to transmit and broadcast high-precision time standard signals. The transmitting antenna adopts the form of four inverted-cone antennas. It is supported by four towers with a height of approximately 206 m. The distance between the sides of the grounding tower is approximately 425 m. Figure 1 shows photos of the four towers of the BPL antenna [1,2].

Figure 1. Four towers of the BPL antenna.

The Changhe II system, which has the form of a single-tower, insulated umbrella tower (see Figure 2), is located in Jilin Helong, Shandong Rongcheng, Anhui Xuancheng, Guangdong Raoping, Guangxi He County, and Chongzuo City. Single-tower umbrella antennas are geometrically symmetric, which can improve the efficiency of the ground net and thus the system transmission efficiency. The horizontal component of the single antenna is small, which reduces the energy loss of the antenna system and the interference of the horizontal component with the waveform [2,3].

2.2. United States

There are many Loran-C transmitting antennas in the USA. Most use tall mast radiators with heights ranging from 190–220 m, which are insulated from the ground. These masts are inductively lengthened and fed by a loading coil. A well-known example of a station that uses such an antenna is Rantum. However, some stations utilize free-standing tower radiators of a similar height (e.g., Carolina Beach). The output power from some Loran-C transmitters exceeds 1000 kW, which is achieved by using extremely tall (412 m) mast radiators. Other high-power Loran-C stations, such as George, use four T-antennas mounted on four guyed masts arranged in a square configuration. Figure 3 shows a single-tower antenna belonging to a radio station in the USA, which is still used for the Loran-C system.

Figure 2. Single umbrella antenna of Changhe II.

Figure 3. Satellite image of a Loran-C system transmitting antenna in the USA.

2.3. Russia

The Chayka system is a Russian ground radio navigation system that shares similarities with the Loran-C system. Chayka transmitters operate at very high power. Certain transmitters use 460 m tall, single-tower mast antenna, while others use multi-tower antennas. In the 1980s, three sets of transmission antennas were constructed as Chayka chains in northern Siberia. These chains were used to transmit 1200 kW navigation signals at 100 kHz and demonstrated high transmission efficiency. Figure 4 shows the 460-m-tall antennas used in the Russian Chayka system.

Figure 4. Satellite image of a Chayka system transmitting antenna in Russia.

3. Structure and Performance Design of Typical Loran-C Transmitting Antenna

3.1. Single-Tower Antenna Form

As the main antenna form of the global Loran-C transmission system, the single tower antenna is usually an umbrella antenna supported by a 200–300 m high tower. The radiator is composed of mast, top line and bottom lead. Part of the top line below the line leads to the lower line at the bottom of the tower feed point. The whole radiator is in a conical umbrella shape. The representative schematic diagram of the structure is shown in Figure 5. The top of the tower is powered by several load top conductors as umbrella conductors, which are connected to the ground through insulators, and the bottom of the mast is connected to the ground through an insulator base. The radiation grounding grid is set with the iron tower as the center, and the ground wire is buried underground. The transmission power of the system is 2000 kW. Table 1 shows the performance indexes of the one-tower transmitting antennas designed with reference to Figure 5 at different heights [8,9].

Table 1. Main Performance Parameters of the Single-Tower Antenna.

The height of Single-Tower Antenna (m)	280	250	200
Antenna efficiency (%)	78.2	77.2	70.61
Input resistance (Ω)	8.09	6.4	3.84
Ground Loss Resistor (Ω)	1.76	1.44	1.13
Static capacitance (nF)	7.57	7.25	5.79
Antenna bandwidth (kHz)	3.84	3.0	1.42
Product of efficiency and bandwidth (kHz)	3.0	2.33	1
Effective height (m)	189	168.1	124.4

The main advantages of the single tower umbrella antenna are that the main body of the antenna system is supported by only one iron tower, the antenna structure is relatively simple, the radiation efficiency is high, the engineering quantity is relatively small, the cost is low and the cost performance is high. Its main disadvantages are that the central tower needs to be insulated from the ground, the performance requirements of the base insulator are very high and the seismic performance of the system is lower than that of the four-tower inverted cone antenna. In addition, in areas prone to snow disasters and sand storms, there may be the risk of system short circuits and work interruption caused by the inundation of the base insulator.

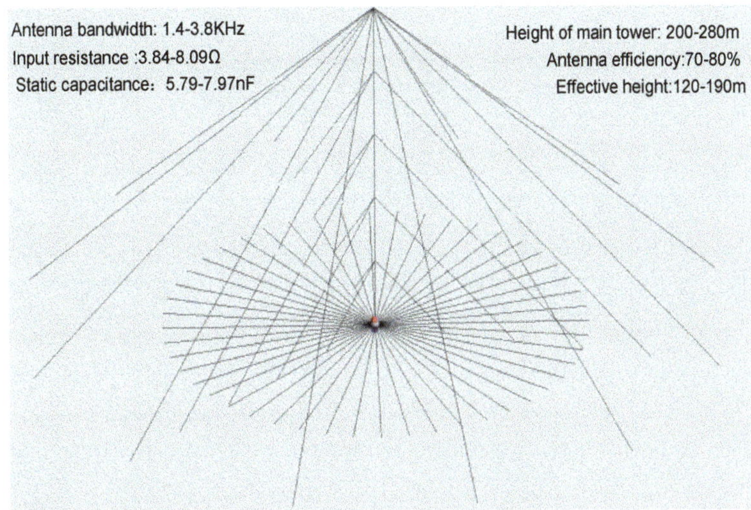

Figure 5. Schematic diagram of a single-tower umbrella antenna model.

3.2. Four-Tower Antenna Form

The antenna is supported by four towers of equal height between 200 and 280 m. The radiating body consists of a square top load, main and lower leads and auxiliary lower leads. A small tower is built at the center of the lower leads to provide fixed closure. A schematic diagram of the four-tower inverted-cone antenna is shown in Figure 6. Owing to the low effective height and radiation resistance of the four-tower inverted-cone antenna, a large-scale ground network needs to be designed to achieve better radiation efficiency. The ground network is radial with the feeding point as its center, which is also buried underground. Table 2 shows the performance indexes of the four-tower transmitting antennas designed with reference to Figure 6 at different heights [10,11].

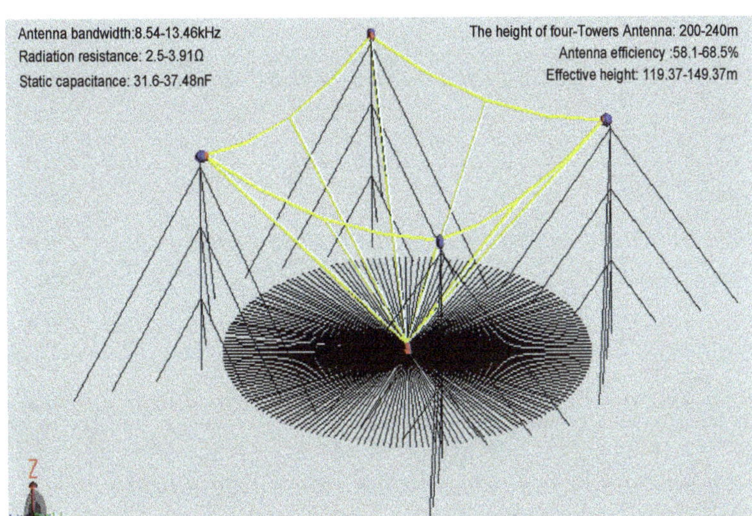

Figure 6. Schematic diagram of the four-tower inverted-cone antenna model.

Table 2. Main Performance Parameters of the four-tower Antenna.

The height of four-Towers Antenna (m)	240	220	200
Tower spacing (m)	507.6	465.3	423
Antenna efficiency (%)	68.5	63.6	58.1
Ground loss resistor (Ω)	1.8	1.8	1.8
Static capacitance (nF)	37.48	34.49	31.6
Radiation resistance (Ω)	3.91	3.14	2.5
Antenna bandwidth (kHz)	13.46	10.72	8.54
Product of efficiency and bandwidth (kHz)	9.22	6.82	4.97
Effective height (m)	149.37	133.88	119.37

The main advantage of the four-tower transmitting antenna is that the tower height is relatively low and the bandwidth is wide. All iron towers are directly grounded to avoid the use of an insulator at the base of the iron tower. In areas with severe environments such as multiple snow disasters and earthquakes, the system has high safety, reliability and good seismic performance. Its main shortcomings are relatively low radiation efficiency, a large number of towers, large quantities and high cost.

4. Antenna Broadcast Signal Propagation Model

According to the atmospheric noise data in the book, *VLF Radio Engineering* of the United States, in Tibet, the maximum atmospheric noise affecting the 100-kHz frequency band occurs during summer, and the maximum noise spectral density is approximately -120 dBV$\cdot$m$^{-1}\cdot$Hz1/2. The receiver bandwidth is 30 kHz (referring to the Loran-C receiver), which corresponds to a noise field strength of 44.7 dB$\cdot\mu$V$\cdot$m^{-1}. Based on the operation experience of the long-running Loran-C navigation and BPL time service systems, a minimum signal field strength of 45 dB$\cdot\mu$V$\cdot$m^{-1} should be maintained to ensure reliable reception [12,13].

With reference to the geographic information data in four directions around the newly-built transmitter, setting the Loran signal propagation model and then simulating the transmission field strength of the antenna with MATLAB tools allows valuable data related to the terrain and signal propagation around the station to be obtained, which can be used to determine the propagation capacity and radiation efficiency requirements of the antenna.

By comparing the structure and electrical performance of single-tower antennas and four-tower antennas, it can be seen that single-tower transmitting antennas have certain performance advantages. In order to further analyze the transmission capacity of a single-tower antenna, the antenna performance can be more directly understood from the aspects of the single-tower transmitting antenna pattern, the field intensity distribution under the antenna, the antenna magnetic field distribution and the radiation efficiency curve. Figure 7 shows the direction diagram of a single-tower antenna, Figure 8 shows the distribution of antenna field strength in the quarter area, Figure 9 shows the distribution of antenna magnetic field in the quarter area and Figure 10 shows the radiation efficiency curve.

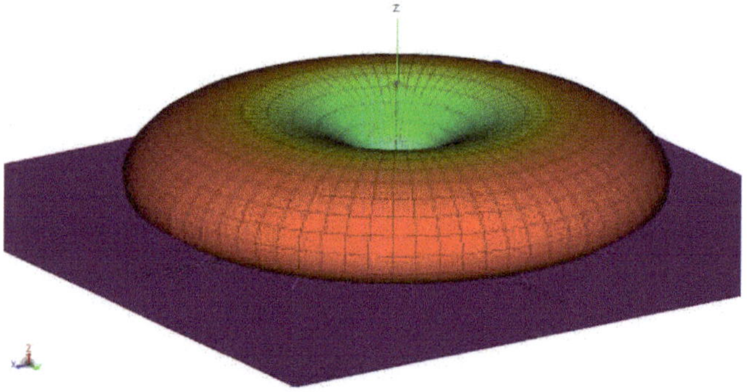

Figure 7. Antenna pattern.

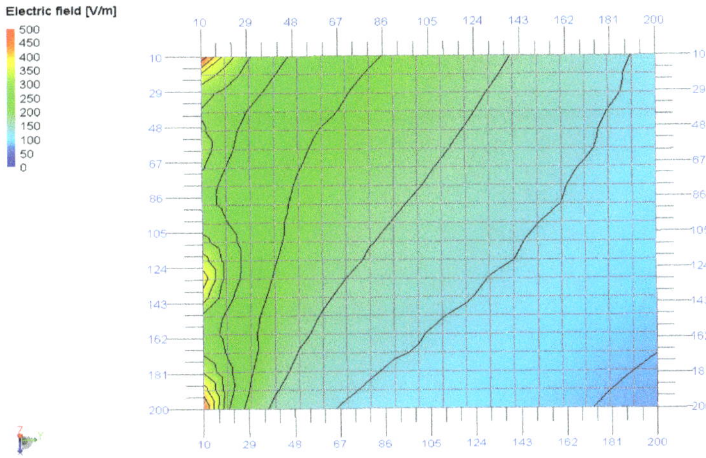

Figure 8. Distribution diagram of ground electric field under antenna (quarter area).

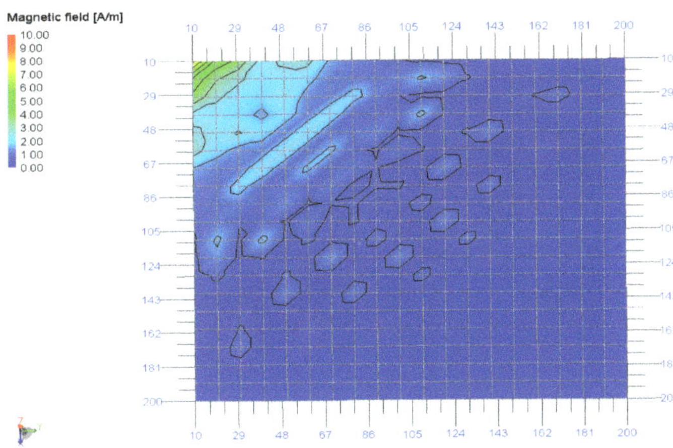

Figure 9. Antenna magnetic field distribution (quarter area).

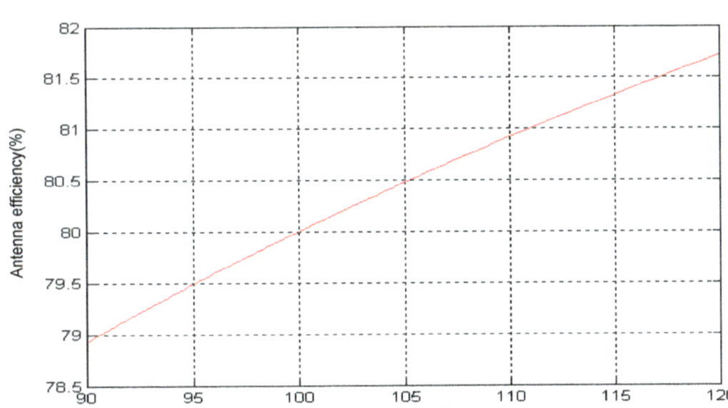

Figure 10. Antenna efficiency curve.

With reference to the geographic information data in four directions around the newly-built transmitter, setting the Loran signal propagation model, and then simulating the transmission field strength of the antenna with MATLAB tools allows valuable data related to the terrain and signal propagation around the station to be obtained, which can be used to determine the propagation capacity and radiation efficiency requirements of the antenna. Therefore, an antenna selection database can be established. Simulating the four-way radio signal propagation around the transmitter in Naqu, Tibet is to finally determine the electrical parameters of the transmitting antenna and better deal with the very special geological and environmental conditions in Naqu, Tibet [14,15].

4.1. Eastward Transmission

During the eastward propagation of the eLoran signal, the field strength of the signal will be affected by the terrain and change accordingly. Simulation Figure 11 shows the change of geographical environment during propagation, while Figure 12 shows that the signal field strength changes with the increase in propagation distance under the influence of geographical environment.

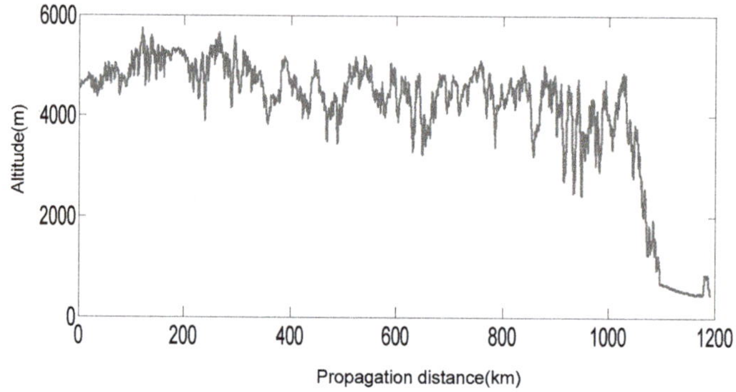

Figure 11. Landform profile of the Naqu region due east.

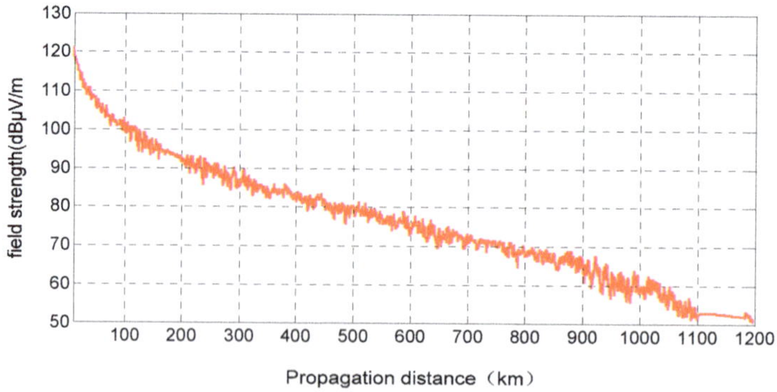

Figure 12. Field intensity propagation curve for the transmitted signal from the eastward-oriented antenna.

4.2. Westward Transmission

During the westward propagation of the eLoran signal, the field strength of the signal will be affected by the terrain and change accordingly. Simulation Figure 13 shows the change of geographical environment in the process of propagation, while Figure 14 shows the change of signal field strength under the influence of geographical environment with the increase in propagation distance.

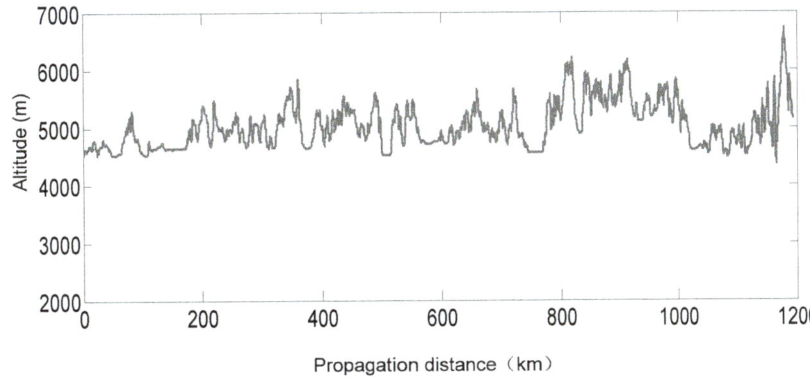

Figure 13. Landform profile of the Naqu region due west.

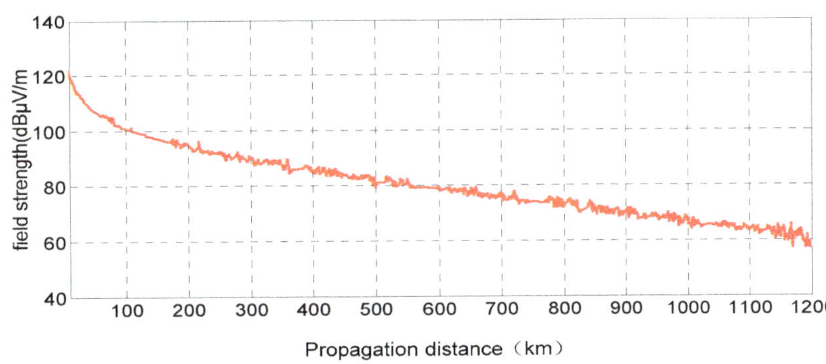

Figure 14. Field intensity propagation curve of the transmitted signal from the westward-oriented antenna.

4.3. Southward Transmission

When the eLoran signal propagates southward, the field strength of the signal will be affected by the terrain and change accordingly. Simulation Figure 15 shows the change of geographical environment during propagation, while Figure 16 shows that the signal field strength changes with the increase in propagation distance under the influence of geographical environment.

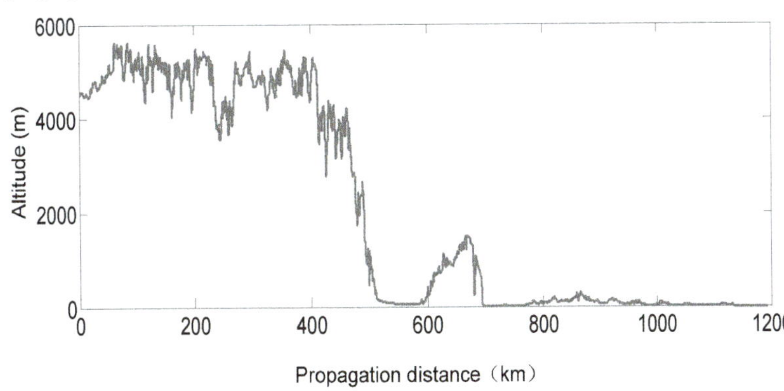

Figure 15. Landform profile of the Naqu region due south.

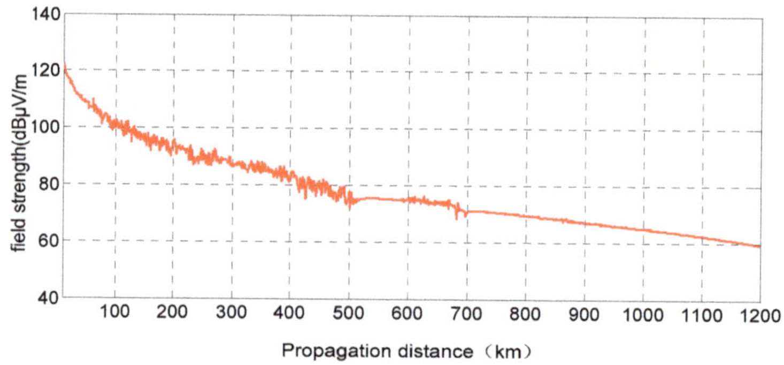

Figure 16. Field intensity propagation curve of the transmitted signal from the southward-oriented antenna.

4.4. Northward Transmission

When the eLoran signal propagates northward, the field strength of the signal will be affected by the terrain and change accordingly. Simulation Figure 17 shows the change of geographical environment during propagation, while Figure 18 shows that the signal field strength changes with the increase in propagation distance under the influence of geographical environment.

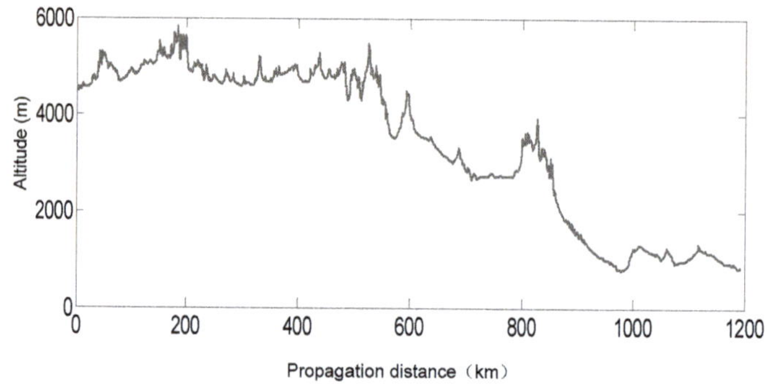

Figure 17. Landform profile of the Naqu region due north.

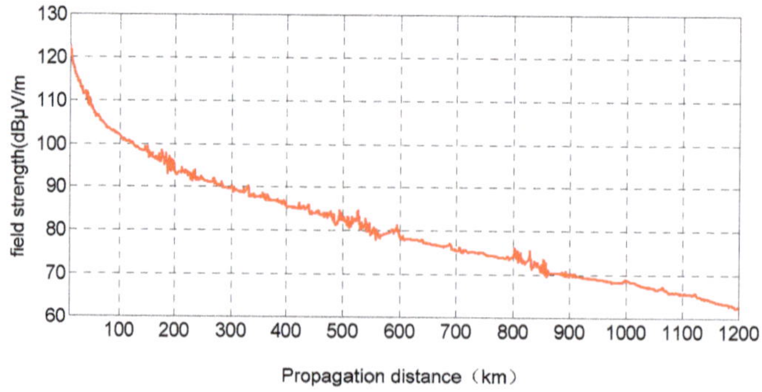

Figure 18. Field intensity propagation curve of the transmitted signal from the northward-oriented antenna.

The simulation results show that the topographic fluctuation of the propagation path has a slight influence on the signal intensity, with the propagation curves showing localized fading characteristics that correspond to the topographic profile. When the topographic fluctuations are less intense, the characteristics of the ground wave propagation decline are not as significant; only a few peaks demonstrate declines measuring 2–3 dB. The results indicate that the radiation signal field intensity of the Naqu transmitter can satisfy the required coverage distance of 1000 km. Furthermore, these results can be used to guide antenna construction in Western China. The results here demonstrate clearly that single-tower antennas with an increased effective height provide the optimal transmission range and efficiency [16–18].

5. Conclusions

Long-wave transmitting antennas have been widely used all over the world. Single-tower or four-tower structures are usually adopted. According to the development status of long-wave, high-power antenna technology at home and abroad, and referring to the practical experience of major Loran transmitting antennas in the world, the performance parameters of various Loran transmitting systems are compared. In addition, there are many innovations in the research work of this paper: firstly, this project will design and construct the Loran transmitting antenna with an altitude of more than 4500 m for the first time in the world. In addition, according to the geographical environment of the Naqu area, the propagation model of the Loran signal is established. Compared with previous papers, the innovations of this paper are shown in the content of this paper to have been rarely carried out before. Compared with the previous similar literature, this paper analyzes and compares the radiation characteristics of the transmitting antenna in detail, and simulates the radiation field strength and signal propagation of the four basic compass directions in the complex geographical environment of Tibet, China, in order to more intuitively design and enhance the signal propagation ability in this area. These simulation results not only show the advantages of the performance of the single-tower antenna in Tibet, but also design the structure and electrical parameters of the transmitting antenna according to the simulation results so that the enhanced Loran signal transmitted by the transmitting antenna can be extended to 1000 km or even further. Therefore, the single-tower umbrella antenna should be selected as the preferred form of enhanced Loran transmitting antenna [19–21].

Author Contributions: Conceptualization, Z.H. and S.L.; methodology, M.D.; software, C.Y.; validation, Z.H., S.L. and C.Y. All authors have read and agreed to the published version of the manuscript.

Funding: This research received no external funding.

Conflicts of Interest: The authors declare no conflict of interest.

References

1. Duan, J. *Study on the Broadcast Control Method of the BPL Long-Wave Time Service Signal*; National Time Service Center: Xi'an, China, 2008.
2. Fang, F.; Wang, Z. Improvement of Some Constructions of Hoisting and Hanging System of BPL Long-wave Transmitting Antenna. *J. Time Freq.* **2012**, *35*, 12.
3. Guo, J.; Xie, H. Research on Influence Supporting Tower to Electrical Properties of Large Long-Wave Antenna. *Mod. Navig.* **2014**, *3*, 201–204.
4. Koo, H.; Nam, S. Modified L-type eloran transmitting antenna for co-location with an AM antenna. In Proceedings of the 2016 International Symposium on Antennas and Propagation (ISAP), Okinawa, Japan, 24–28 October 2016; pp. 370–371.
5. Sun, S.; Li, Y. Electrical Characteristics of VLF Umbrella Antenna. *Ship Electron. Eng.* **2016**, *6*, 59–60.
6. Xu, Y.; Geng, J.; Zhuang, K.; Wu, H.; Han, J.; Wang, K.; Zhou, H.; Jin, R.; Lian, X. Study on miniaturized super low frequency(SLF) transmitting antenna. In Proceedings of the 2019 13th European Conference on Antennas and Propagation (EuCAP), Krakow, Poland, 31 March–5 April 2019; pp. 1–3.
7. Liang, M.; Zhang, H.-Y.; Zhang, F.-S.; Sun, F.-K. A novel single-fed high-gain and phase-adjustable transmitting antenna element for wireless power transmission. In Proceedings of the 2018 IEEE MTT-S International Wireless Symposium (IWS), Chengdu, China, 6–9 May 2018; pp. 1–4. [CrossRef]
8. Papamichael, V.C.; Karadimas, P. Performance Evaluation of Actual Multielement Antenna Systems Under Transmit Antenna Selection/Maximal Ratio Combining. *IEEE Antennas Wirel. Propag. Lett.* **2011**, *10*, 690–692. [CrossRef]

9. Kawakami, H.; Haga, T.; Hosoi, K.; Shirahama, D.; Norimatsu, Y.; Ninomiya, Y.; Tanioka, M. Digital terrestrial broadcasting antennas -4-Plane synthesis pattern and gain improvement-. In Proceedings of the 2007 IEEE Antennas and Propagation Society International Symposium, Honolulu, HI, USA, 9–15 June 2007; pp. 4721–4724. [CrossRef]
10. Erfani, E.; Tatu, S.-O.; Niroo-Jazi, M.; Safavi-Naeini, S. A millimeter-wave transmitarray antenna. In Proceedings of the 2016 17th International Symposium on Antenna Technology and Applied Electromagnetics (ANTEM), Montreal, QC, Canada, 10–13 July 2016; pp. 1–2. [CrossRef]
11. Hansen, P.M.; Rodriguez, A. Performance analysis of large electrically small transmit antennas. In Proceedings of the 2011 IEEE International Symposium on Antennas and Propagation (APSURSI), Spokane, WA, USA, 3–8 July 2011; pp. 782–785. [CrossRef]
12. Li, H.; Liu, C. Calculation on characteristics of VLF umbrella inverted-cone transmitting antenna. In Proceedings of the 2014 Sixth International Conference on Ubiquitous and Future Networks (ICUFN), Shanghai, China, 8–11 July 2014; pp. 389–391. [CrossRef]
13. Monin, A. Submarine floating antenna model for LORAN-C signal processing. *IEEE Trans. Aerosp. Electron. Syst.* **2003**, *39*, 1304–1315. [CrossRef]
14. Bian, Y.; Last, J. Loran-C skywave delay estimation using eigen-decomposition techniques. *Electron. Lett.* **1995**, *31*, 133–134. [CrossRef]
15. Johler, J. The propagation time of a radio pulse. *IRE Trans. Antennas Propag.* **1963**, *11*, 661–668. [CrossRef]
16. Wu, H.; Li, X.; Zhang, H.; Gao, H.; Bian, Y. UTC message broadcasting over Loran-C data channel. In Proceedings of the 2002 IEEE International Frequency Control Symposium and PDA Exhibition (Cat. No.02CH37234), New Orleans, LA, USA, 29–31 May 2002; pp. 530–536. [CrossRef]
17. Tsao, C.; Debettencourt, J. Measurement of the Phase Constant for Rock Propagated Radio Signals. *IEEE Trans. Commun. Technol.* **1967**, *15*, 592–597. [CrossRef]
18. Gressang, R. Estimating Bias in Loran Lines of Position. *IEEE Trans. Aerosp. Electron. Syst.* **1970**, *6*, 400–405. [CrossRef]
19. Yuguo, C.; Panpan, B.; Shuji, H. A novel polarization measurement method for large transmitting antenna/antenna array. In Proceedings of the 2016 11th International Symposium on Antennas, Propagation and EM Theory (ISAPE), Guilin, China, 18–21 October 2016; pp. 175–178. [CrossRef]
20. Uno, T.; Adachi, S. Range distance requirements for large antenna measurements. *IEEE Trans. Antennas Propag.* **1989**, *37*, 707–720. [CrossRef]
21. Kim, S. Efficient Transmit Antenna Selection for Receive Spatial Modulation-Based Massive MIMO. *IEEE Access* **2020**, *8*, 152034–152044. [CrossRef]

applied sciences

MDPI

Article

A Modified Compact Flexible Vivaldi Antenna Array Design for Microwave Breast Cancer Detection

Ayman M. Qashlan, Rabah W. Aldhaheri * and Khalid H. Alharbi

Department of Electrical and Computer Engineering, King Abdulaziz University, Jeddah 21589, Saudi Arabia; agashlan1991@gmail.com (A.M.Q.); khalharbi@kau.edu.sa (K.H.A.)
* Correspondence: raldhaheri@kau.edu.sa

Abstract: In this paper, a compact, flexible Vivaldi antenna is designed, and an array of nine identical antennas of this type is used as a microwave breast imaging model to detect cancerous tumors in the multilayers phantom model presented in this paper. The nine-antenna array is used to measure the backscattering signal of the breast phantom, where one antenna acts as a transmitter and the other eight antennas act as receivers of the scattered signals. Then, the second antenna is used as a transmitter and the other antennas as receivers, and so on till we have gone through all the antennas. These collected backscattered signals are used to reconstruct the image of the breast phantom using software called "Microwave Radar-based Imaging Toolbox (MERIT)". From the reconstructed image, the tumor inside the breast model can be identified and located. Different tumor sizes in different locations are tested, and it is found that the locations can be determined irrespective of the tumor size. The proposed modified Vivaldi antenna has a very compact size of $25 \times 20 \times 0.1$ mm^3 and has a different geometry compared with conventional Vivaldi antennas. The first version of the antenna has two resonant frequencies at 4 and 9.4 GHz, and because we are interested more in the first band, where it gives us sufficient resolution, we have notched the second frequency by etching two slots in the ground plane of the antenna and adding two rectangular parasitic elements on the radiating side of the antenna. This technique is utilized to block the second frequency at 9.4 GHz, and, as a result, the bandwidth of the first resonant frequency is enhanced by 20% compared with the first design bandwidth. The modified antenna is fabricated on Polyimide flexible material 0.1 mm thick with a dielectric constant of 3.5 using a standard PCB manufacturing process. The measured performance of this antenna is compared with the simulated results using the commercially available simulation software Ansoft HFSS, and it is found that the measured results and the simulated results are in good agreement.

Keywords: flexible antenna; Vivaldi antenna; breast imaging; microwave imaging; cancer detection

Citation: Qashlan, A.M.; Aldhaheri, R.W.; Alharbi, K.H. A Modified Compact Flexible Vivaldi Antenna Array Design for Microwave Breast Cancer Detection. *Appl. Sci.* **2022**, *12*, 4908. https://doi.org/10.3390/app12104908

Academic Editors: Naser Ojaroudi Parchin and Ernesto Limiti

Received: 11 April 2022
Accepted: 10 May 2022
Published: 12 May 2022

1. Introduction

Breast cancer is considered a major health issue across the globe. It is considered the most common cancer that affects women among all cancer cases. In 2022, an estimated 287,850 new cases of invasive breast cancer will be diagnosed in the U.S. alone, which contributes to 30% of all cancer cases among women [1–3]. The early detection of such malignant cells is one of the most significant factors in improving the survival rate and quality of life experienced by breast cancer sufferers. X-ray mammography is the current detection method for early-stage breast cancer. However, this method is an invasive and ionizing technique. It also yields high false-negative rates [4,5]. Other non-ionizing detection methods, such as Magnetic Resonance Imaging (MRI), can be used for early-stage breast cancer diagnoses. However, this method is expensive and may not be accessible to a large number of patients [5,6].

In recent years, there has been a great demand for a new reliable, non-ionizing, cost-effective, and comfortable approach to breast cancer screening. Microwave imaging (MWI)

methods are one of the promising techniques for early breast cancer diagnosis [7]. At microwave frequencies, the contrast between the electrical properties of cancerous cells and those of healthy breast tissue is significant. Several experimental studies of dielectric properties of healthy and malignant breast tissues have been reported in the literature. In [8], a large-scale study, with 319 measurements on freshly excised breast tissue specimens to experimentally determine the ultrawideband microwave dielectric properties of a variety of normal, malignant, and benign breast tissues, measured from 0.5 to 20 GHz, is reported. The analysis showed that the contrast in dielectric properties between malignant and normal adipose-dominated tissues in the breast is considerable, as large as 10:1. Moreover, the contrast in the microwave-frequency dielectric properties between malignant and normal glandular/fibroconnective tissues in the breast is no more than about 10%. In [9], experimental measurements of dielectric properties of normal and tumorous tissues on more than 220 tissue samples were performed in the frequency range of 0.5 to 50 GHz. The results showed that the mean values for tumorous and normal tissues are separated from each other at all frequencies. Moreover, the normal tissue samples were divided into three subgroups based on their adipose component percentage. It has been found that as fat content decreases, the average dielectric properties rise significantly between groups. An average contrast in dielectric properties of malignant and healthy tissues of 8.4:1, 2.2:1, and 1.4:1 exist for low-adipose, medium-adipose, and high-adipose tissues, respectively. Similar experimental results with 330 samples were also reported in [10]. This reported contrast allows for smaller breast tumors detection with higher accuracy than traditional methods [11,12].

In microwave imaging, there are two approaches: tomography- and radar-based. In microwave tomography, the dielectric properties of the breast are calculated by transmitting narrowband microwave signals through the breast. The backscattered signals are collected through multiple receiving antennas. However, this method requires many transmitting and receiving antennas. Additionally, it requires solving a non-linear inverse scattering problem which requires a computationally intensive reconstruction algorithm. In radar-based imaging, antennas are used to transmit and receive the backscattered signals using the transmitting antenna (monostatic) or additional antennas (multistatic). The monostatic arrangement can be used multiple times from multiple locations to provide a sufficient number of channels for imaging. Alternatively, the multistatic antenna array method uses multiple antennas at a fixed location to collect the backscattered signals. This method avoids the mechanical issues of monostatic imaging systems. Successful imaging systems based on this idea have been developed in [13–15].

In MWI, antennas play a vital role in the imaging process. They act as transceivers where a transmitting antenna becomes a receiving antenna (sensor) in the next iteration. Thus, careful antenna design is required to meet system requirements. Recent studies have shown that antennas used for MWI should have the following properties: small size, high gain, directive power radiation, and wide bandwidth. The operational frequency band is also important: lower frequency bands provide greater penetration, and higher bands offer better range resolution. However, lower frequency requires a larger antenna footprint, and higher frequency causes higher power losses [16,17].

UWB antennas have attracted researchers for their unique features: high data rate, small size, low cost, and power spectrum density. They can also be designed to operate in both low and high-frequency ranges. Moreover, they are environmentally friendly, biocompatible, and biologically friendly [18]. Several types of UWB antenna for MWI have been proposed, including omnidirectional vs. directional radiation pattern [19]; wide range vs. narrowband [20]; high vs. low frequency [21]; etc. However, in all cases, the system requires high efficiency, high gain, and compatibility to penetrate the human body [3]. To date, many UWB antennas for MWI breast imaging have been reported: compact planar UWB antenna [22], modified antipodal Vivaldi antenna [3], hemispherical antenna [17], reflectarray antenna [23], flexible monopole antenna [7], circular polarized radial line slot array antenna [24], and many more.

The human breast, on the other hand, has an inhomogeneous and complex structure of skin, fat, gland, and muscle. Each of these substances has different dielectric properties, namely relative permittivity and conductivity. This inhomogeneous environment can be modeled by stacking several homogeneous layers that represent the properties of skin, fat, gland, and muscle. The proposed structure is not a representation of the real breast. However, the model is widely used in laboratory environments for prototype breast imaging systems [5,7,25,26].

Successful imaging systems require preprocessing for backscattered signals to remove artefacts and reflections. Several artefact removal algorithms have been reported in the literature. The Average Subtraction method subtracts the response of each channel from a reference waveform. The reference waveform is the average response of all imaging channels [27]. This method assumes that the artefact is similar in each channel. Adaptive filtering extends this idea to compensate for channel-to-channel variation in the artefact [28]. The rotational subtraction method has been reported in [29]. In this method, the antenna array is physically rotated around its center, and a second radar measurement is performed. The two data sets are then subtracted to remove undesired signals.

In this paper, a modified and compact flexible Vivaldi antenna with a size of $25 \times 20 \times 0.1$ mm^3 is proposed and studied. Two versions of antennas, one with two resonant frequencies at 4 and 9.4 GHz and the other with one resonant frequency at 4.4 GHz, are discussed. Both are of the same size, and the improved antenna is fabricated on Polyimide flexible material with a thickness of 0.1 mm and a dielectric constant of 3.5 using a standard PCB manufacturing process. The reflection coefficients, current distribution, and radiation patterns are discussed and evaluated. An array of nine identical antennas of the second version, i.e., with one resonant frequency at 4.4 GHz, is used as a microwave breast imaging model to detect and locate the tumor inside the proposed multilayer phantom model using Microwave Radar-Based Imaging toolbox (MERIT). This simulation setup successfully identifies and locates tumors of different sizes and positions inside the phantom model. The imaging results are very promising for the practical use of the proposed microwave imaging as a good candidate for breast cancer detection. The novelty of the work is the use of the proposed modified flexible antenna and the proposed breast phantom for breast imaging applications.

2. Materials and Methods

2.1. Antenna 1 Design

A simple, low-cost, compact, and flexible Vivaldi UWB antenna is designed. The antenna has a radiating triangular shape on the top side and a slotted ground plane on the bottom side. It has a relatively small size of 25×20 mm^2 compared with conventional Vivaldi antennas presented in the literature. It is fabricated using standard printed circuit board (PCB) manufacturing processes with polyimide flexible material. To the best of our knowledge, the compact size of the flexible Vivaldi antenna presented here can be considered one of the smallest sizes that have been reported in the literature. Figure 1 illustrates the geometry of the proposed flexible Vivaldi antenna 1. The antenna is fabricated on an inexpensive Polyamide substrate with a thickness of 0.1 mm, relative permittivity of 3.5, and loss tangent, $\tan\delta = 0.02$. The radiating side of the antenna on the top layer consists of a microstrip feed line, a horizontal line that passes above the parabolic notch on the GND plane, and a triangular shape to enhance propagation. The feeder line of the antenna consists of two segments of the line where the first segment has a width of 0.21 mm, which results in a 50 Ω matching impedance. The second segment of the feeder has a width of 0.28 mm to match the first segment of the feeder to the horizontal line of the radiation side. The horizontal line extends through the triangular shape of the radiating side of the antenna. The antenna has a 50 Ω CPW feeder line on the top side, and an SMA connector is connected at the end of the line. The proposed antenna underwent many design stages. Starting with a $45 \times 40 \times 1.5$ mm^3 Vivaldi antenna, the parameters were optimized to yield the proposed final shape of $25 \times 20 \times 0.1$ mm^3. The radiating shape of the antenna was also investigated to yield the best possible outcome. The designed dimensions of the proposed Vivaldi antenna were obtained using the commercially available simulation software Ansoft HFSS. The optimized parameters of the final design of proposed Antenna 1 are given in Table 1.

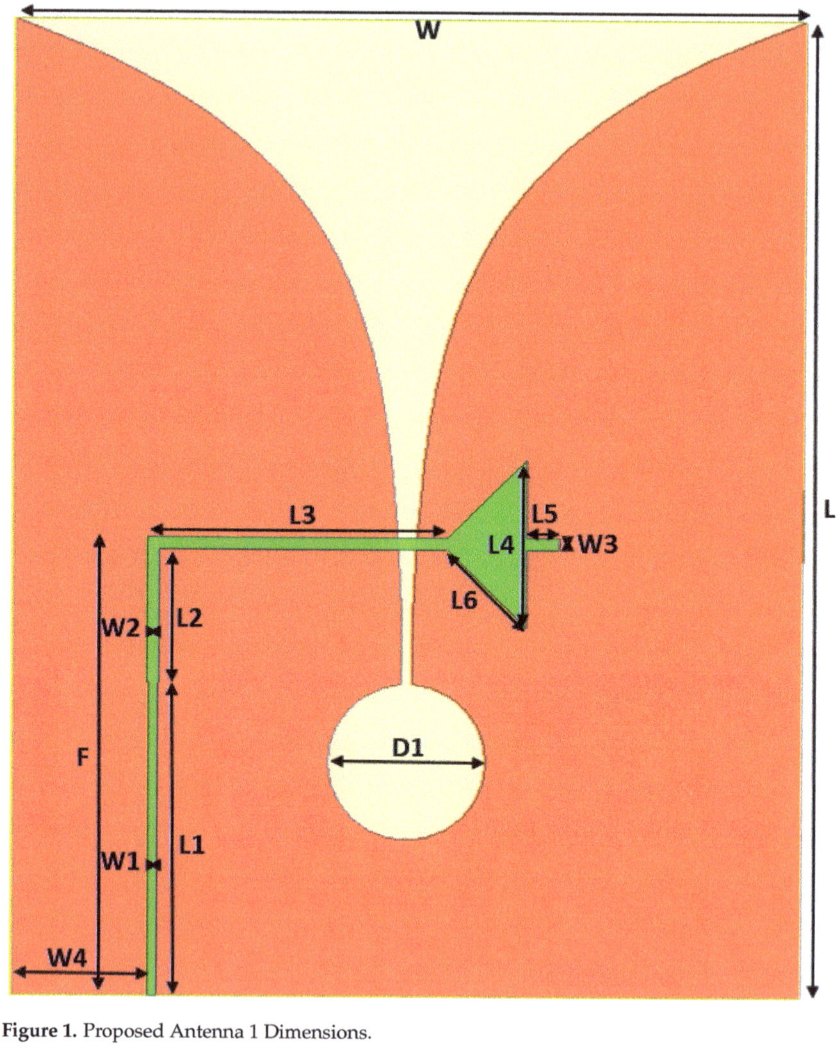

Figure 1. Proposed Antenna 1 Dimensions.

Table 1. Proposed Optimal Dimensions of Antenna 1.

Parameter	Value (mm)	Parameter	Value (mm)
L	25	W	20
L1	8	L2	3.45
L3	7.54	L4	4.27
L5	0.8	L6	2.83
W1	0.21	W2	0.28
W3	0.27	D1	4
W4	3.5	F	11.45

The simulation results show that the antenna has two resonant frequencies: the first resonant frequency is at 4.0 GHz with a 1 GHz impedance bandwidth, and the second is located at 9.4 GHz with a 1 GHz impedance bandwidth. Figure 2 shows the reflection coefficient of this antenna against frequency.

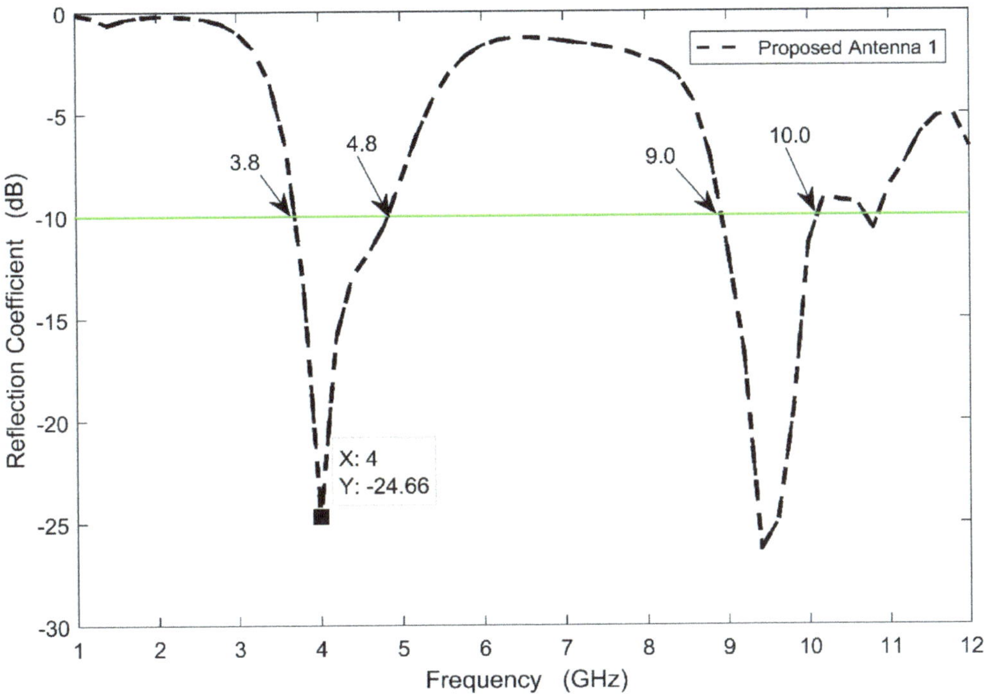

Figure 2. Reflection Coefficient of Proposed Antenna 1.

2.2. Modified Antenna with One Resonant

Proposed Antenna 1 has two resonant frequencies, at 4 GHz and 9.4 GHz. However, only the bandwidth of the first resonant frequency is of interest. In this section, a modified antenna is presented to notch the second resonant frequency and further enhance the bandwidth at the first resonant frequency. The modified antenna has the same dimensions as Proposed Antenna 1. However, two slots (DGS) in the ground planes were added to notch the second resonant frequency at 9.4 GHz. The DGS creates a notch frequency at 8.2 GHz, which eliminates the second resonant frequency. Moreover, copper rectangles were added to the propagator side of the antenna. The rectangles work as parasitic elements through which the bandwidth of the antenna is enhanced. Modified Antenna dimensions are shown in Figure 3. The modified antenna has the same dimensions as Proposed Antenna 1 except for the added slots and parasitic elements. The dimension parameters of the added elements are listed in Table 2.

Figure 4 shows the reflection coefficients of the modified antenna. As can be seen, the modified antenna has only one resonant frequency, at 4.4 GHz. Moreover, the impedance bandwidth of the modified antenna, 1.2 GHz, is greater than the bandwidth of Proposed Antenna 1 by 20%. The current distribution of the modified antenna is higher at the feeder line of the radiator. On the bottom side, the current is concentrated around the edges of the parabolic cut in the ground plane, with a higher current around DGS, as shown in Figure 5. The gain, as demonstrated in Figure 6, increases over frequency, and it is at its minimum at the notch frequency. It has a local maximum of 2.33 dBi at 4.4 GHz, the center frequency of the antenna. The gain is at its minimum at 6.7 GHz, which corresponds to the maximum value of the reflection coefficient shown in Figure 4. The modified antenna has a directive radiation pattern. Figure 7 shows the 3D radiation pattern of the modified antenna. The E-plane and H-plane are shown in Figure 8, where the XZ plane represents the E-plane, and the YZ plane represents the H-plane.

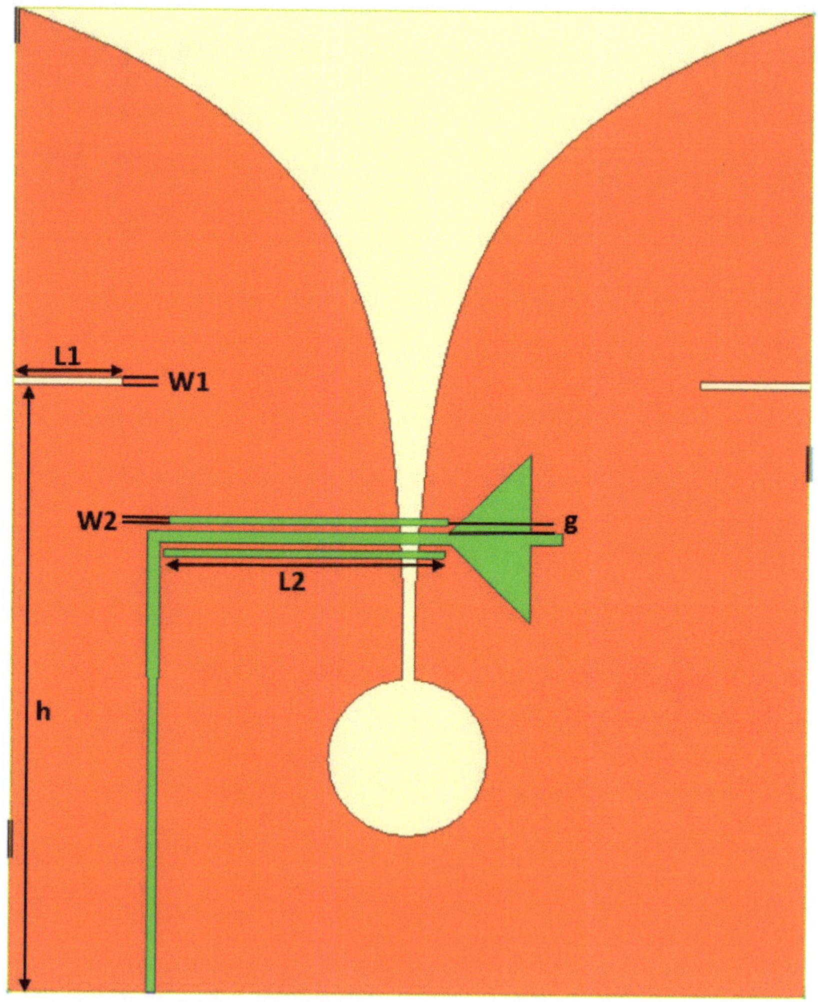

Figure 3. Modified Antenna's dimensions.

Table 2. Modified Antenna optimal parameters.

Parameter	Value (mm)	Parameter	Value (mm)
L1	2.8	L2	7
W1	0.2	W2	0.15
h	15.2	g	0.2

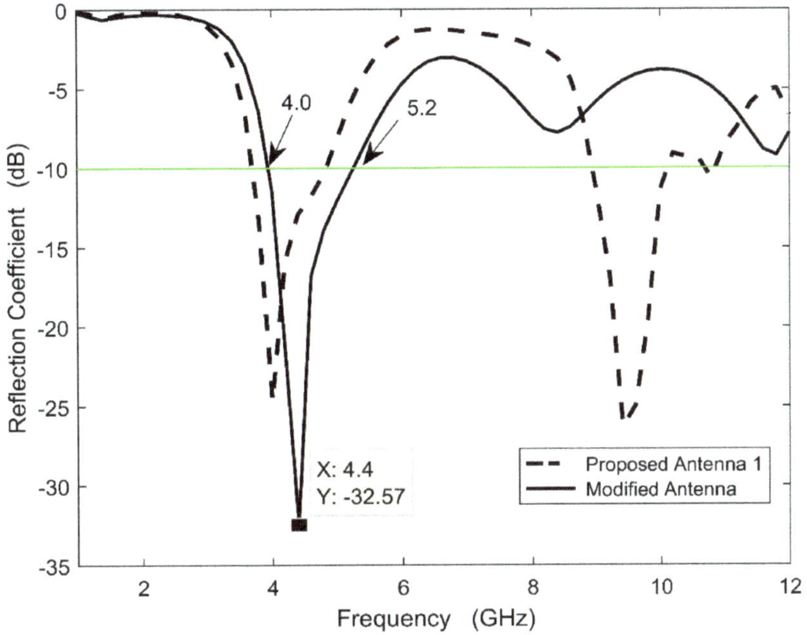

Figure 4. Reflection Coefficient of the Modified Antenna.

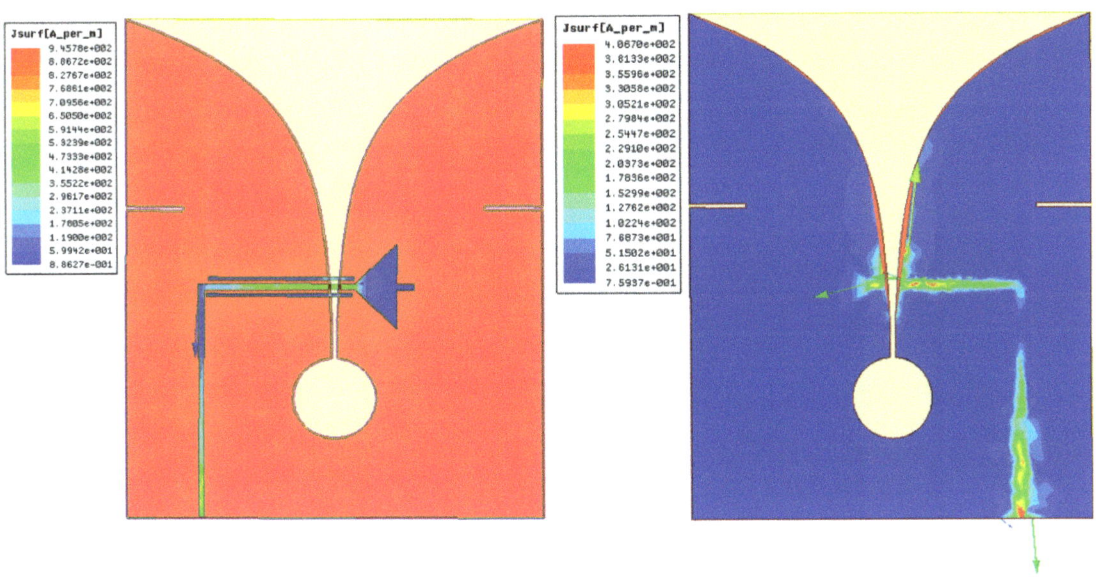

(**a**) (**b**)

Figure 5. Current Distribution at 4.3 GHz: (**a**) Top Side; (**b**) Bottom Side.

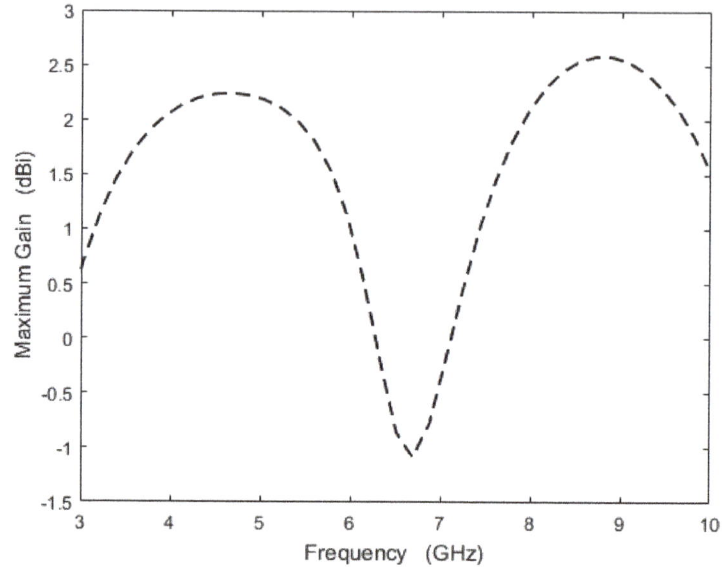

Figure 6. Maximum Gain of the Modified Antenna.

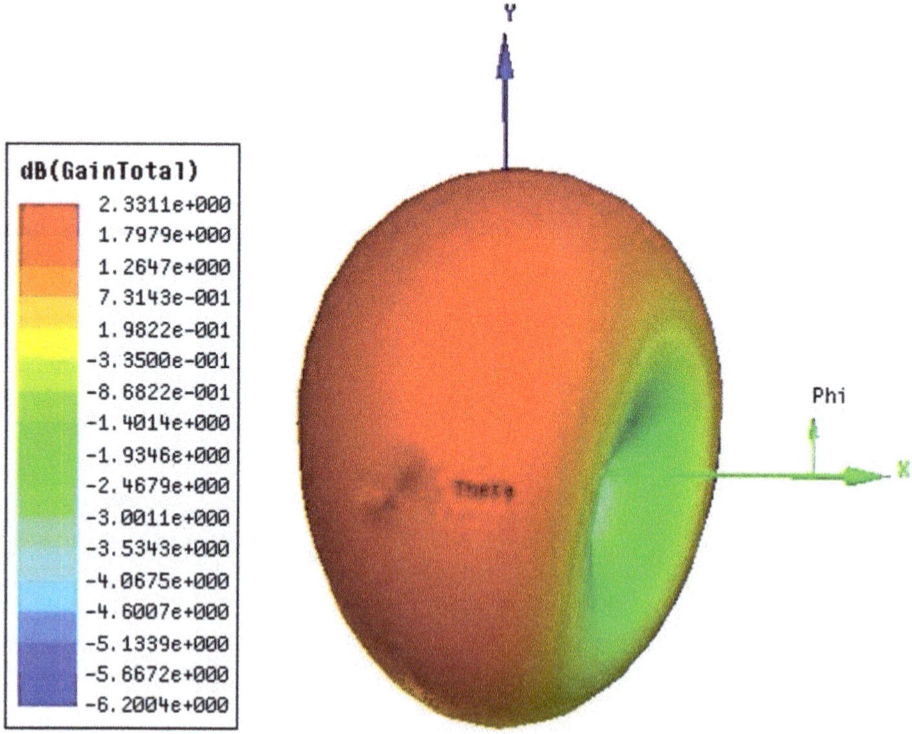

Figure 7. Three-dimensional Radiation Pattern of the Modified Antenna at 4.3 GHz.

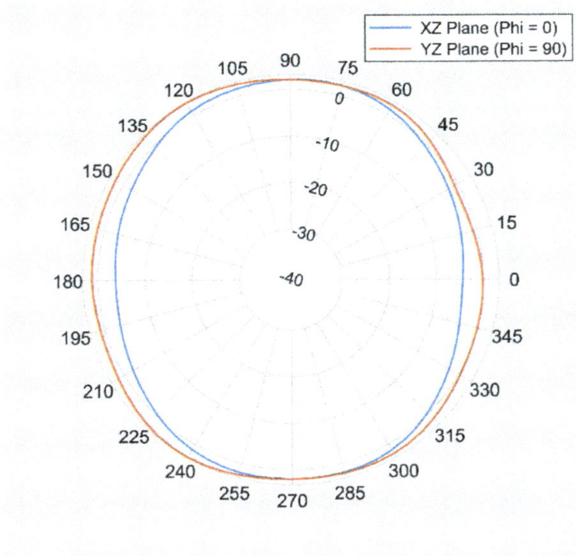

Figure 8. Two-dimensional Radiation Pattern at 4 GHz of Modified Antenna.

2.3. Parametric Study of Modified Antenna

To achieve the highest possible bandwidth of the antenna and enhance the reflection coefficient, the DGS parameters and location were varied. Moreover, the dimensions and size of the parasitic element were studied. The modified antenna achieves the best possible result. The studies detailed below show the effects of changing each parameter.

2.3.1. Changing DGS Width, W1

The DGS width affects the bandwidth at the first and second resonant frequency. As can be seen in Figure 9, W1 = 0.2 mm shows the best result.

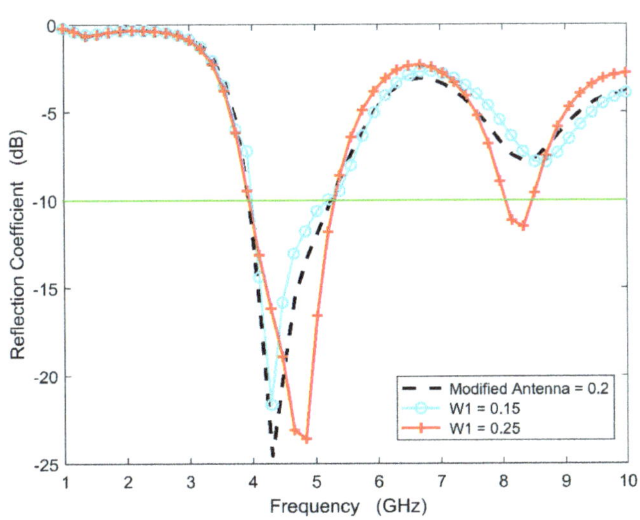

Figure 9. Effects of Changing W1.

2.3.2. Changing Parasitic Element Width, W2

Reducing W2 results in a slightly higher BW but increases the reflection coefficient. It also introduces a second resonant at 8.2 GHz. Increasing W2 enhance the reflection coefficient but reduces the bandwidth. Figure 10 shows the effects of changing W2. It can be seen that W2 = 0.15 mm is the optimum value.

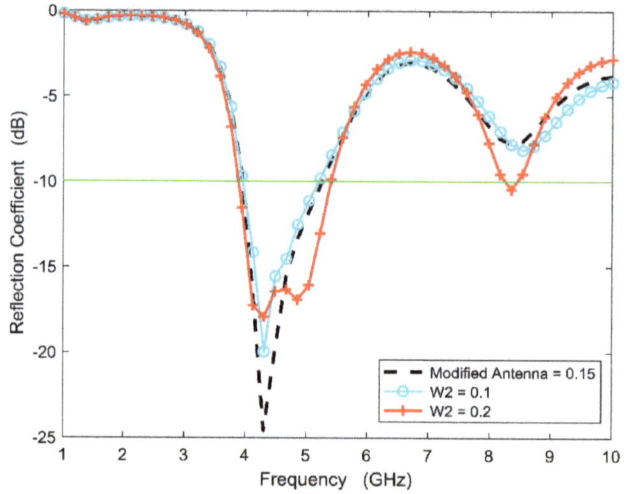

Figure 10. Effects of Changing W2.

2.3.3. Changing DGS Length, L1

The length of DGS affects the bandwidth of the antenna at 4.4 GHz, as shown in Figure 11. A smaller value of L1 slightly decreases the BW of the antenna and enhances the reflection coefficient of the antenna. A larger value of L1 enhances the bandwidth but greatly increases the reflection coefficient.

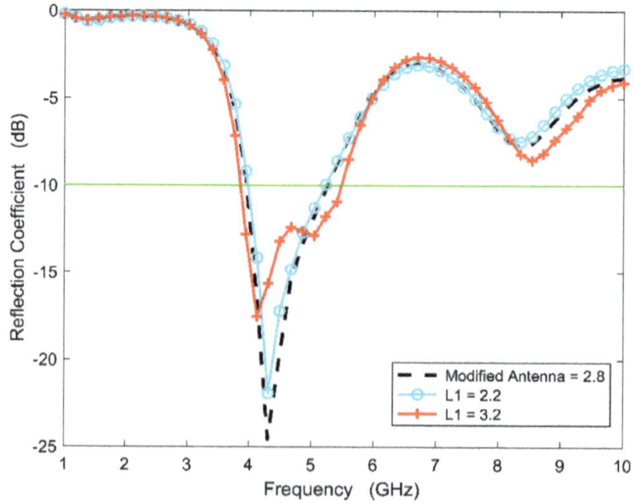

Figure 11. Effects of DGS Length L1.

2.3.4. Changing Parasitic Element Length, L2

Increasing L2 results in a similar BW but also increases the reflection coefficient, as shown in Figure 12. Reducing L2 increases the BW but introduces a second resonant frequency at 8.8 GHz and increases the reflection coefficient.

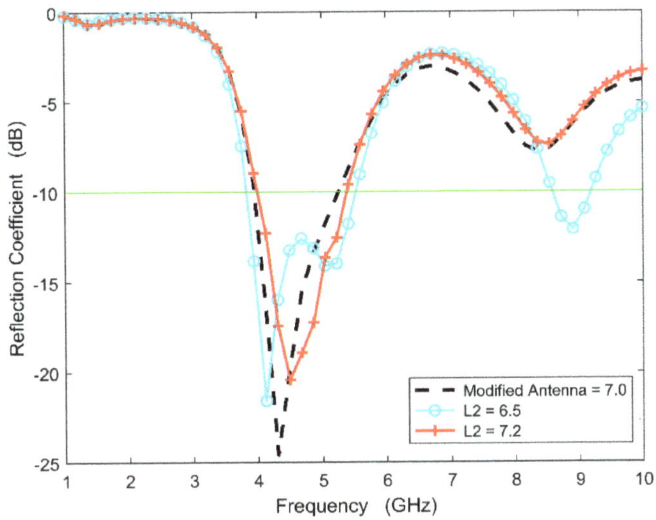

Figure 12. Effects of Changing L2.

2.3.5. Changing DGS Height, h

The height of the DGS affects the resonant frequency and BW of the antenna, as shown in Figure 13. A higher value of h results in a better BW but introduces a second resonant frequency at 8.2 GHz, which conflicts with the purpose of the modified antenna. On the other hand, a smaller value of h results in a higher reflection coefficient.

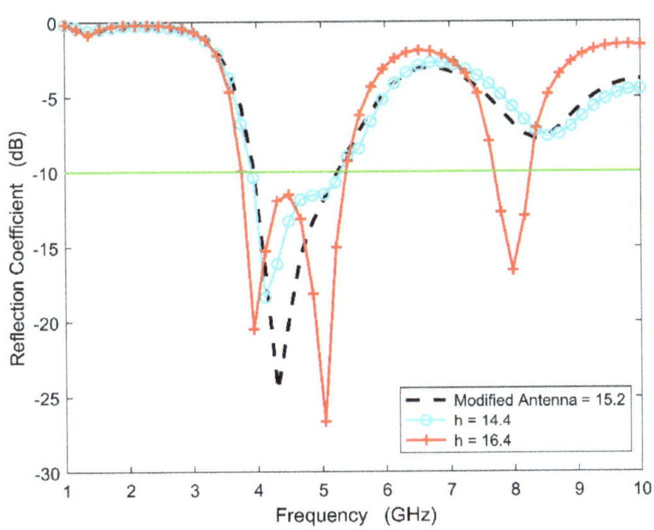

Figure 13. Effects of Changing DGS Height.

2.3.6. Changing Parasitic Element Gap, g

The gap between the parasitic element and the radiator affects the center frequency location and reflection coefficient values. As shown in Figure 14, g = 0.2 mm is the optimum value for this antenna.

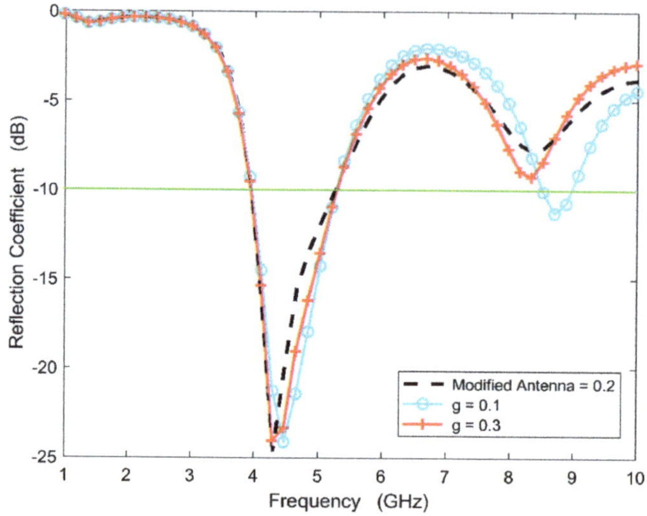

Figure 14. Effects of Changing Parasitic Element Gap.

2.4. Comparison to Literature

The proposed antenna was compared with various antennas reported in the literature. The results are summarized in Table 3. The proposed antenna has a smaller BW than most reported antennas. However, a BW greater than 1 GHz is considered sufficient for breast imaging. Moreover, the proposed antenna reduces the size of a conventional Vivaldi antenna by half while keeping a directional radiation pattern.

Table 3. Comparison of Proposed Antennas and others reported in the Literature.

Citation	Type	Radiation	Size mm^3	Band (GHz)	Fc (GHz)	BW (GHz)	Gain (dBi)
[3]	Vivaldi	Directional	40 × 40 × 1.6	2.5–11	NA	8.5	7.2
[7]	Flexible Monopole	Omni-directional	20 × 20 × 0.05	2–4	3	2	NA
[30]	Vivaldi	Directional	36 × 36 × 1.6	3–12	7.2	9	8.2
[31]	Vivaldi	Directional	48 × 46 × 0.8	3.1–10.6	7.8	7.5	8.25
[32]	Vivaldi	Directional	57 × 41 × 1.6	3–9	3.8	6	NA
[33]	Vivaldi	Directional	51 × 42 × 0.05	2.8–7	5	4.2	7.5
Proposed 1 With two resonant	Vivaldi	Directional	25 × 20 × 0.1	3.8–4.8 and 9–10	4.0 and 9.5	1.0 and1.0	2.24 and 2.7
Proposed 2 With one resonant	Vivaldi	Directional	25 × 20 × 0.1	4.0–5.2	4.4	1.2	2.33

2.5. Antenna Manufacture

The modified antenna was manufactured on flexible polyimide material with 18 μm copper from both sides using standard PCB manufacturing processes. A standard SMA connector was attached to the antenna as a feeder. The connector has four ground pins and is suitable for a PCB thickness of 1.6 mm. To overcome this issue, two ground pins were removed from the connector, and the connector was assembled diagonally so that it fit without bending the antenna. Figure 15 shows the top and bottom sides of the manufactured antenna after assembly. The reflection coefficient of the fabricated antenna was measured using Vector Network Analyzer (VNA). The result showed that the resonant frequency was shifted to 4.0 GHz. Moreover, the measured reflection coefficient showed that the antenna has higher bandwidth than the simulated antenna. The actual antenna has a bandwidth of 1.7 GHz centered at 4 GHz. The difference in antenna bandwidth and center frequency is mainly due to the fabrication and measurement errors. Figure 16 shows simulated and measured reflection coefficients. The radiation pattern of the antenna was measured in an anechoic chamber. Figure 17 shows the measured and simulated radiation patterns for the E-plane and H-plane at 4.3 GHz. The results are close to simulation results except for manufacturing and assembly errors.

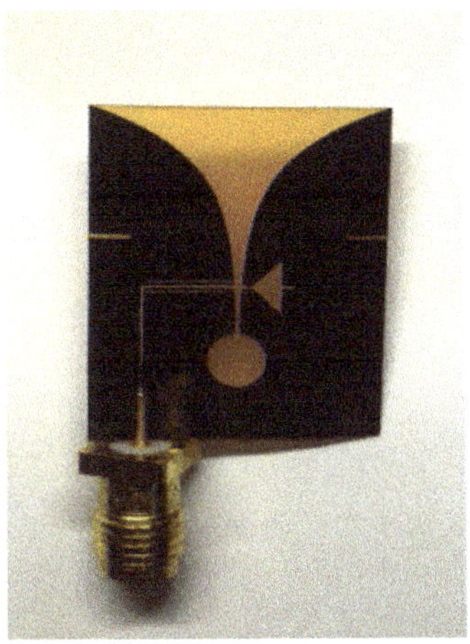

(a)

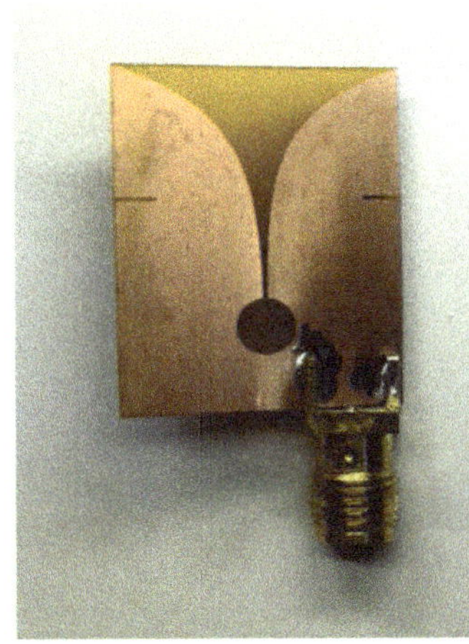

(b)

Figure 15. Modified Antenna: (**a**) Top Side; (**b**) Bottom Side.

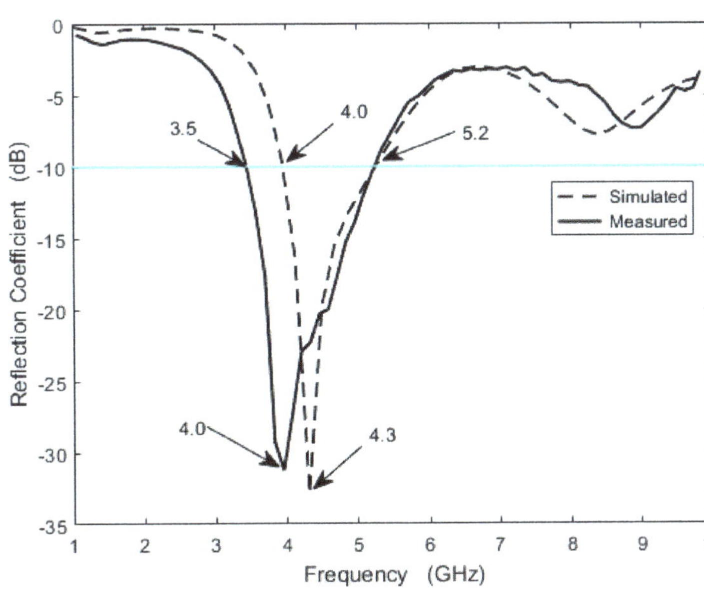

Figure 16. Simulated and Measured Reflection Coefficients.

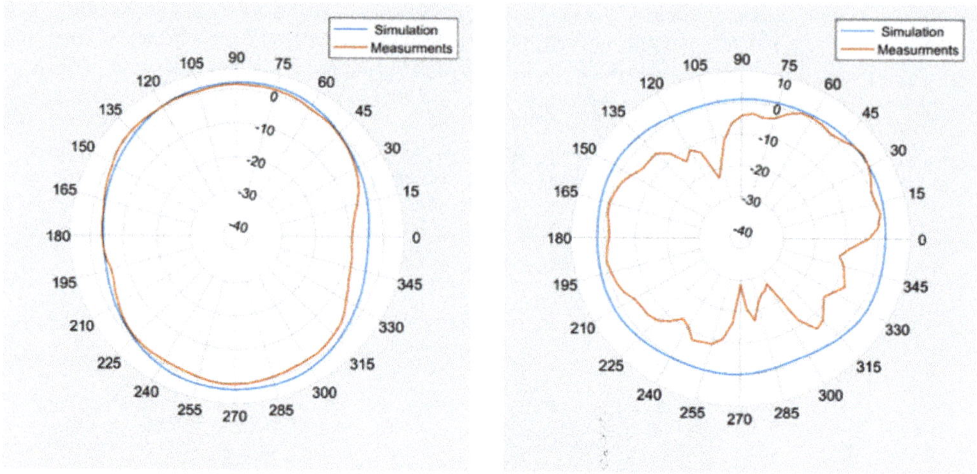

(**a**) (**b**)

Figure 17. Radiation Pattern at 4.3 GHz: (**a**) E-Plane; (**b**) H-Plane.

2.6. Breast Model

A large-scale study of microwave dielectric properties of normal, benign, and malignant breast tissue was reported in [8–10]. It has been found that normal breast tissue spans a very large range of dielectric properties dependent on the adipose content of the sample. Low water content (high adipose) exhibits a low dielectric constant, and high water content glandular or fibroconnective tissue exhibits a high dielectric constant. In this paper, a half-spherical breast phantom was designed with electrical properties that match the human breast. The phantom radius is 45 mm, and it consists of three layers. The skin layer width is 2.5 mm with a dielectric constant and conductivity of 36 and 4 S/m, respectively. The width of the fat (tissue) layer is 42.5 mm with a dielectric constant and conductivity of 9 and 0.4 S/m, respectively. A spherical tumor of 2.5 mm and 5 mm radius is inserted inside the phantom in different places. The tumor has a relatively high dielectric constant of 55 and conductivity of 4 S/m. The proposed phantom presented in our work has similar dielectric properties and thicknesses to phantoms found in [3,14,20]. The proposed model is developed in HFSS, as shown in Figure 18.

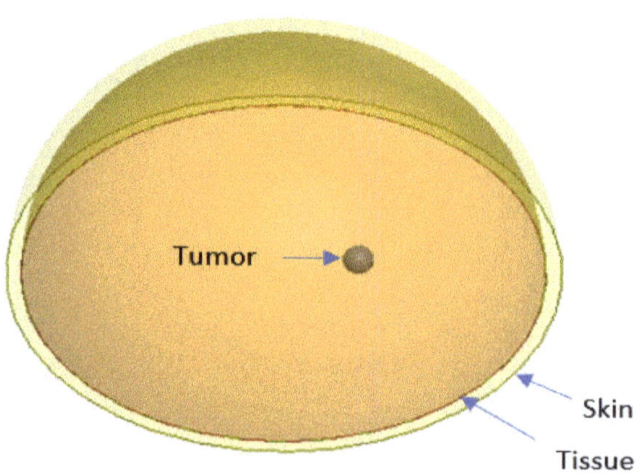

Figure 18. Proposed Breast Phantom.

2.7. Imaging Setup

The imaging setup consists of the proposed breast phantom surrounded by transmitting/receiving antenna elements. The number of antennas used for imaging affects the accuracy of the reconstructed image. In this paper, the modified Vivaldi antenna is used as a radiating/receiving element. Nine antenna elements are used as radiating/receiving elements. The placement of the antenna around the breast phantom is critical in this design since the modified antenna has a directional radiation pattern. The nine antennas are placed at an equal distance from each other and 10 mm from the phantom, with the radiating side of the antenna facing the phantom, as shown in Figure 19. The distance from the center of the phantom to the center of each antenna element is r_3 = 67.5 mm, and the distance between adjacent antennas in the array is 46.2 mm (center to center). The setup is illustrated in Figure 19 with antennas numbered Antenna1–Antenna9. The fidelity of the modified antenna in the imaging setup was investigated. A signal centered at 4.3 GHz is transmitted from Antenna1 and received by all other antennas. The procedure is repeated nine times with Antenna2, Antenna3, ... and Antenna9 as transmitting elements and all others as receiving elements. Backscattered signals are collected in the frequency range of 3.65 to 5.2 GHz. Table 4 shows the minimum reflected signal magnitude of S12, S13, ... , S21, S23, ... , S97, S98 with and without the presence of the tumor inside the breast phantom. The result shows that the presence of the tumor increases the reflected signal magnitude due to the presence of high-dielectric material. Moreover, a weaker signal magnitude is received when the tumor is not present in the breast phantom. The presence of tumor is very noticeable over the frequency range of 4.1 to 4.7 GHz. The imaging is performed in air; a coupling medium between the array and the phantom is not used.

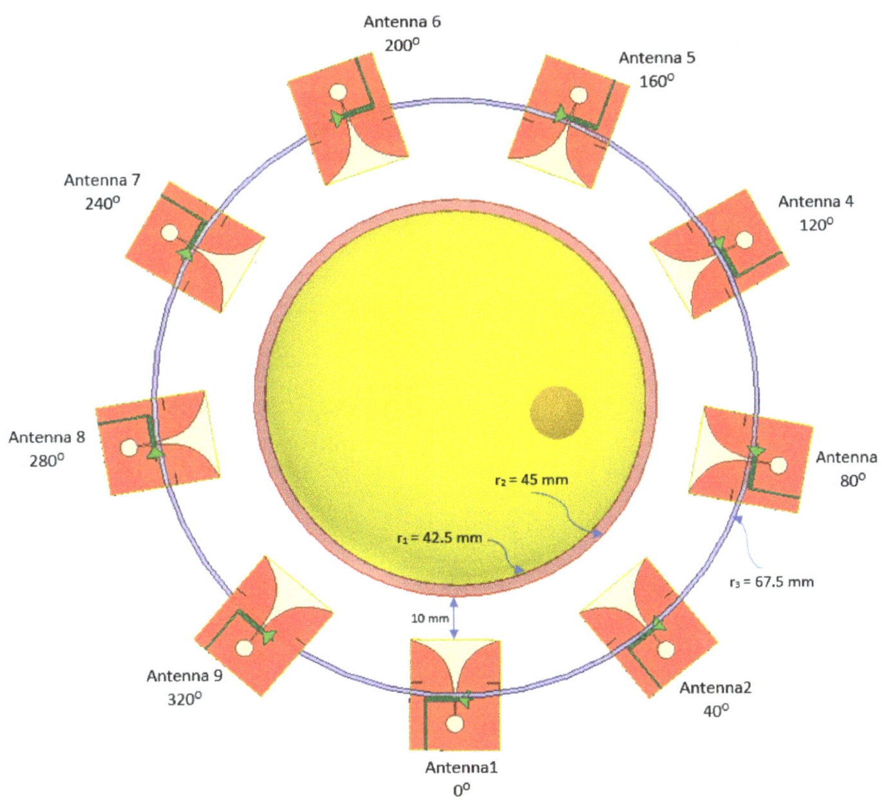

Figure 19. Imaging Setup using Modified Antenna.

Table 4. Backscattered Signal Magnitude with and without the Presence of Tumor.

Frequency Range (GHz)	Minimum Reflected-Signal Magnitude without Tumor (dB)	Minimum Reflected-Signal Magnitude with Tumor (dB)	Variation in Magnitude (dB)	Presence of Tumor
3.65–3.8	−7.5	−5.6	−1.9	Weak Noticeable
3.8–3.95	−10.6	−7.7	−2.9	Weak Noticeable
3.95–4.11	−16.1	−11.9	−4.2	Noticeable
4.1–4.25	−30.3	−21.8	−8.5	Strong Noticeable
4.25–4.4	−34.0	−18.6	−15.5	Strong Noticeable
4.4–4.6	−28.0	−16.4	−11.6	Strong Noticeable
4.6–4.7	−21.5	−14.3	−7.2	Strong Noticeable
4.7–4.9	−16.4	−11.7	−4.7	Noticeable
4.9–5	−14.0	−10.6	−3.4	Noticeable
5–5.2	−16.5	−13.0	−3.5	Weak Noticeable

2.8. Image Reconstruction

Image reconstruction was performed using open-source Matlab software that was developed by Martin Glavin, Edward Jones, and Martin O'Halloran. The software is called "The Microwave Radar-based Imaging Toolbox (MERIT) [34]. It aims to produce a robust, consistent framework for microwave image signal processing and reconstruction.

In the imaging software, collected signals are preprocessed to remove artefacts from system components such as skin and tissues, as well as antenna coupling signals. The software uses the Rotational Subtraction method to perform artefact removal. The imaging setup shown in Figure 19 is used to perform the first radar measurement set. After that, the antenna array is physically rotated around its center by 36°, and a second radar measurement set is performed. Undesired signals such as those from skin, tissues, and antenna coupling are almost identical and appear at the same time position. In contrast, tumor response appears at a different time position in the two data sets. Thus, the imaged object (tumor) response is isolated, and clutter is eliminated by subtracting the rotated scan from the original scan. Applying this technique depends on the homogeneity of the breast within the rotation angle. Therefore, during antenna array rotation, the distance between antennas and skin remains unchanged, while skin and tissue properties and thickness are the same.

The filtered signals are used for image reconstruction using the Delay-and-Sum (DAS) method. The method is based on synthetically focusing signals on points within the imaging domain. Synthetically focused signals from each channel are summed. Then, the energy of the summed signals is used to reconstruct the energy profile of the imaging domain. Points of high contrast exhibit coherent addition from multiple channels resulting in high energy at that point; thus, the areas of high dielectric contrast are highlighted. Synthetic focusing is achieved by compensating attenuation and phase for a given channel. The procedure is summarized in the steps below:

- Propagation path is estimated from the Euclidian distance between the transmitting antenna, the points of interest, and the receiving antenna;
- Distance traveled through different media along the path is calculated;
- Dielectric properties of the media are estimated based on published dielectric datasets.

The distance travelled and the dielectric properties are used to synthetically focus the signals.

3. Results

The proposed imaging setup shown in Figure 19 was used to detect tumors at different locations inside the proposed half-spherical breast model. The imaging is performed in the commercially available software HFSS in the frequency range from 4.0 GHz to 5.2 GHz. Moreover, imaging is performed with a radius of 45 mm and a resolution of 5×10^{-4}. As shown in the figures below, the proposed system could detect a small tumor of a 2.5 mm radius. Figures 20–24 show the imaging setup and reconstructed image. The reconstructed image represents the breast phantom where the center of the phantom is the (0,0) point of the reconstructed image. The image extends from −45 to 45 mm on both x- and y-axis, which represent the 2d surface of the breast. The image has a color scale where objects in yellow have a high dielectric constant, indicating tumors. The dark blue object indicates normal healthy tissues. The red circle in the reconstructed image shows the actual size and location of the tumor. Table 5 summarizes tumor locations and sizes with respect to the breast phantom center. As can be seen, the proposed system efficiently detected the tumors with the correct location and size.

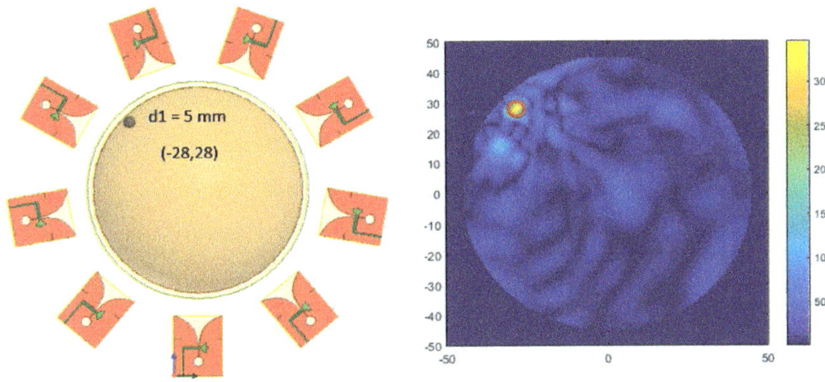

(a) (b)

Figure 20. Tumor Detection: (**a**) Imaging Setup; (**b**) Detected Tumor.

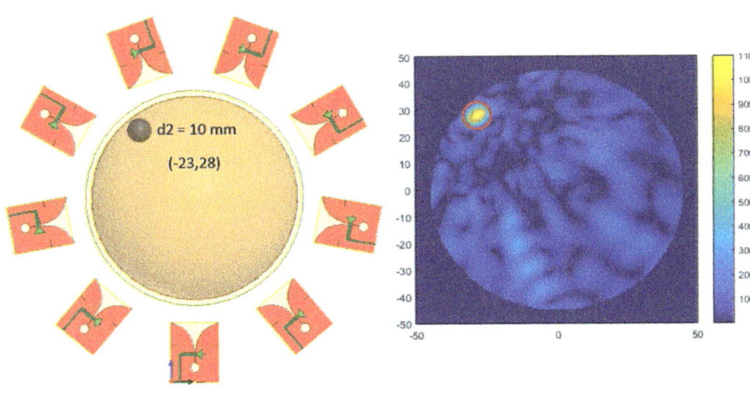

(a) (b)

Figure 21. Tumor Detection: (**a**) Imaging Setup; (**b**) Detected Tumor.

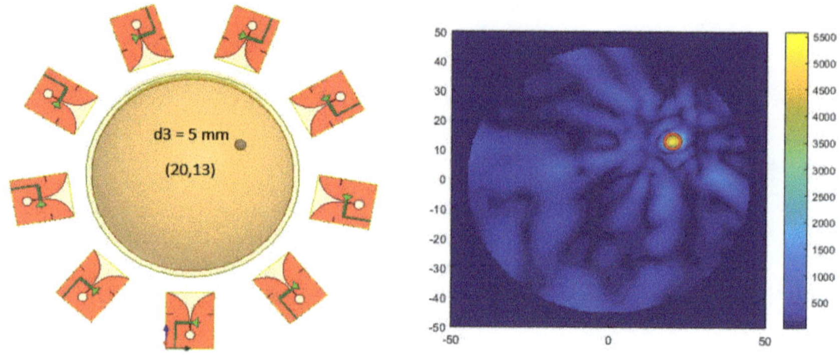

(a) (b)

Figure 22. Tumor Detection: (**a**) Imaging Setup; (**b**) Detected Tumor.

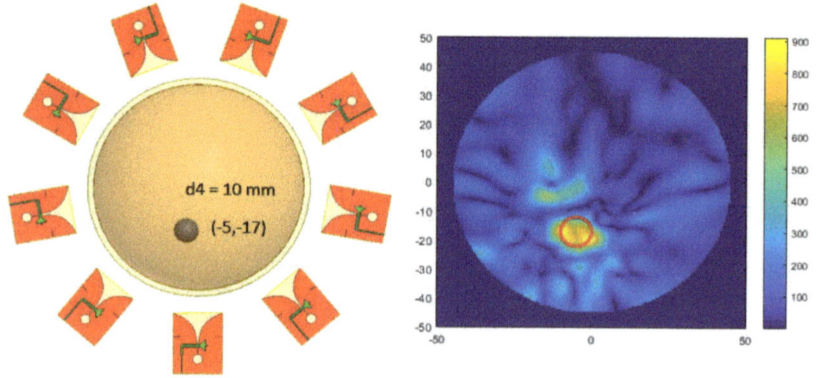

(a) (b)

Figure 23. Tumor Detection: (**a**) Imaging Setup; (**b**) Detected Tumor.

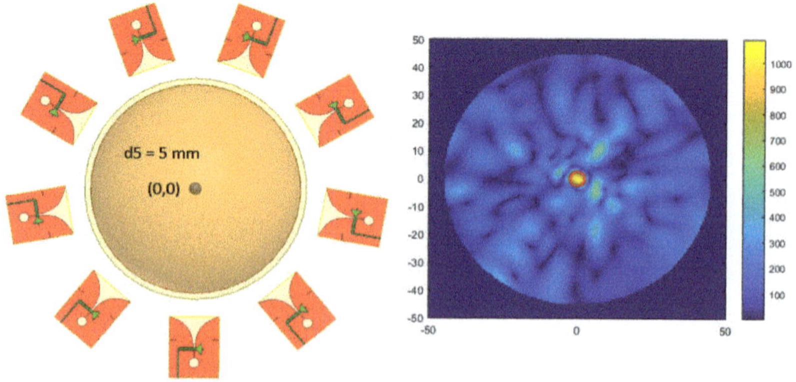

(a) (b)

Figure 24. Tumor Detection: (**a**) Imaging Setup; (**b**) Detected Tumor.

Table 5. Tumor Locations and Sizes.

Imaging Setup	Location (x,y)	Tumor Size (mm) (Diameter)
Figure 20	−28, 28	5
Figure 21	−23, 28	10
Figure 22	20, 13	5
Figure 23	−5, −17	10
Figure 24	0, 0	5

4. Discussion

Tumors of different sizes located at different locations inside the phantom were detected. The proposed system shows great accuracy at distinguishing cancerous masses inside human breast. Tumors were detected near the skin layer, as shown in Figures 20 and 21. Although the skin layer has a relatively high dielectric constant of 36, the image reconstruction algorithm was capable of clearly identifying cancerous masses. In Figures 22 and 23, tumors were detected further away from the skin layer inside the tissue layer. The system was also capable of detecting small tumors of size 5 mm and higher. The worst-case scenario is when the tumor is located at the furthest point inside the phantom, the center. In this setup, the initial radar measurement set and the rotated measurement set are similar, which complicates the detection. Figure 24 shows a 5 mm tumor located at (0,0), where it is clearly identified from healthy cells. However, in all cases, an acceptable amount of clutter is present in the reconstructed images due to the limited bandwidth of the antenna.

5. Conclusions

In this paper, a compact flexible Vivaldi antenna for breast imaging application is proposed. The resultant antenna has two resonant frequencies, at 4 and 9.4 GHz. To eliminate the second frequency and enhance the bandwidth of the proposed Vivaldi antenna, two slots have been etched on the ground side of the antenna. Moreover, two rectangular parasitic elements in the radiation side of the antenna have been added. Due to the added DGS and parasitic elements, the resonant frequency is shifted to 4.4 GHz and the simulated bandwidth increased by 20%. The antenna is fabricated on Polyimide flexible material with a thickness of 0.1 mm and dielectric constant of 3.5 using standard PCB manufacturing processes.

The measured and the simulated reflection coefficient are compared, and it is found that they are in good agreement with each other, as discussed in Section 2.5. An array of 9 antennas is used as a microwave breast imaging model to detect and locate the tumor inside the proposed multilayer phantom model using Microwave Radar Based Imaging toolbox (MERIT). This simulation setup successfully identifies and locates the tumors of different sizes and positions inside the phantom model. A more complex breast phantom model can be used that includes more layers such as the fibroglandural tissue, with a dielectric constant close to that of tumor. Moreover, an imaging setup with a higher number of antennas may be used to enhance imaging quality. In the nine-antenna array, the resolution we obtained is good enough to detect a tumor of 5 mm in diameter, but if we consider more antennas, we might be able to detect tumors of smaller sizes. This will be considered in a future study to generate a more realistic model.

Author Contributions: Conceptualization, A.M.Q. and R.W.A.; methodology, A.M.Q., R.W.A., and K.H.A.; software, A.M.Q.; validation, A.M.Q., R.W.A., and K.H.A.; formal analysis, A.M.Q.; investigation, A.M.Q.; writing—original draft preparation, A.M.Q.; writing—review and editing, R.W.A.; visualization, A.M.Q.; supervision, R.W.A.; project administration, R.W.A. All authors have read and agreed to the published version of the manuscript.

Funding: This research was funded by the Deanship of Scientific Research (DSR) at King Abdulaziz University, Jeddah, Saudi Arabia, under grant number FP-208-43.

Institutional Review Board Statement: Not applicable.

Informed Consent Statement: Not applicable.

Data Availability Statement: Not applicable.

Conflicts of Interest: The authors declare no conflict of interest.

References

1. Siegel, R.L.; Miller, K.D.; Fuchs, H.E.; Jemal, A. Cancer statistics. *CA Cancer J. Clin.* **2022**. [CrossRef]
2. Byrne, D.; Ohalloran, M.; Jones, E.; Glavin, M. A comparison of data-independent microwave beamforming algorithms for the early detection of breast cancer. In Proceedings of the Annual International Conference of the IEEE Engineering in Medicine and Biology Society, Minneapolis, MN, USA, 3–6 September 2009. [CrossRef]
3. Samsuzzaman, M.; Islam, M.T.; Shovon, A.; Faruque, R.I.; Misran, N. A 16-modified antipodal Vivaldi Antenna Array for microwave-based breast tumor imaging applications. *Microw. Opt. Technol. Lett.* **2019**, *61*, 2110–2118. [CrossRef]
4. Munawar, A.; Adabi, S.; Ismail, A.; Saripan, M.; Mahmood, R.; Mahadi, W.; Abdullah, R. Breast cancer detection using Forward Scattering Radar technique. In Proceedings of the IEEE International RF and Microwave Conference, Kuala Lumpur, Malaysia, 2–4 December 2008. [CrossRef]
5. Li, Q.; Xiao, X.; Wang, L.; Song, H.; Kono, H.; Liu, P.; Lu, H.; Kikkawa, T. Direct Extraction of Tumor Response Based on Ensemble Empirical Mode Decomposition for Image Reconstruction of Early Breast Cancer Detection by UWB. *IEEE Trans. Biomed. Circuits Syst.* **2015**, *9*, 710–724. [CrossRef] [PubMed]
6. Woten, A.; Lusth, J.; El-Shenawee, M. Interpreting Artificial Neural Networks for Microwave Detection of Breast Cancer. *IEEE Microw. Wirel. Compon. Lett.* **2007**, *17*, 825–827. [CrossRef]
7. Bahrami, H.; Porter, E.; Santorelli, A.; Gosselin, B.; Popovic, M.; Rusch, L.A. Flexible sixteen monopole antenna array for microwave breast cancer detection. In Proceedings of the 36th Annual International Conference of the IEEE Engineering in Medicine and Biology Society, Chicago, IL, USA, 26–30 August 2014. [CrossRef]
8. Lazebnik, M.; Popovic, D.; McCartney, L.; Watkins, C.B.; Lindstrom, M.J.; Harter, J.; Sewall, S.; Ogilvie, T.; Magliocco, A.; Breslin, T.M.; et al. A large-scale study of the ultrawideband microwave dielectric properties of normal, benign and malignant breast tissues obtained from cancer surgeries. *Phys. Med. Biol.* **2007**, *52*, 6093. [CrossRef]
9. Martellosio, A.; Bellomi, M.; Pasian, M.; Bozzi, M.; Perregrini, L.; Mazzanti, A.; Svelto, F.; Summers, P.E.; Renne, G.; Preda, L. Dielectric properties characterization from 0.5 to 50 GHz of breast cancer tissues. *IEEE Trans. Microw. Theory Tech.* **2017**, *65*, 998–1011. [CrossRef]
10. Di Meo, S.; Espin-Lopez, P.F.; Martellosio, A.; Pasian, M.; Bozzi, M.; Perregrini, L.; Mazzanti, A.; Svelto, F.; Summers, P.E.; Renne, G.; et al. Dielectric properties of breast tissues: Experimental results up to 50 GHz. In Proceedings of the 12th European Conference on Antennas and Propagation, London, UK, 9–13 April 2018. [CrossRef]
11. Misilmani, H.M.; Naous, T.; Khatib, S.K.; Kabalan, K.Y. A survey on antenna designs for breast cancer detection using microwave imaging. *IEEE Access* **2020**, *8*, 102570–102594. [CrossRef]
12. Woten, D.A.; El-Shenawee, M. Broadband Dual Linear Polarized Antenna for Statistical Detection of Breast Cancer. *IEEE Trans. Antennas Propag.* **2008**, *56*, 3576–3580. [CrossRef]
13. Sugitani, T.; Kubota, S.; Toya, A.; Xiao, X.; Kikkawa, T. A Compact 4 × 4 Planar UWB Antenna Array for 3-D Breast Cancer Detection. *IEEE Antennas Wirel. Propag. Lett.* **2013**, *12*, 733–736. [CrossRef]
14. Ouerghi, K.; Fadlallah, N.; Smida, A.; Ghayoula, R.; Fattahi, J.; Boulejfen, N. Circular antenna array design for breast cancer detection. In Proceedings of the Sensors Networks Smart and Emerging Technologies (SENSET), Beiriut, Lebanon, 12–14 September 2017. [CrossRef]
15. Alibakhshikenari, M.; Virdee, B.S.; Shukla, P.; Parchin, N.O.; Azpilicueta, L.; See, C.H.; Abd-Alhameed, R.A.; Falcone, F.; Huynen, I.; Denidni, T.A.; et al. Metamaterial-inspired antenna array for application in microwave breast imaging systems for tumor detection. *IEEE Access* **2020**, *8*, 174667–174678. [CrossRef]
16. Jalilvand, M.; Li, X.; Zwirello, L.; Zwick, T. Ultra wideband compact near-field imaging system for breast cancer detection. *IET Microw. Antennas Propag.* **2015**, *9*, 1009–1014. [CrossRef]
17. Craddock, I.J.; Klemm, M.; Leendertz, J.; Preece, A.W.; Benjamin, R. An improved hemispherical antenna array design for breast imaging. In Proceedings of the 2nd European Conference on Antennas and Propagation, Edinburgh, UK, 11–16 November 2007. [CrossRef]
18. Mahmud, M.; Islam, M.; Misran, N.; Almutairi, A.; Cho, M. Ultra-Wideband (UWB) Antenna Sensor Based Microwave Breast Imaging: A Review. *Sensors* **2018**, *18*, 2951. [CrossRef] [PubMed]
19. Mobashsher, A.T.; Abbosh, A.M. Performance of Directional and Omnidirectional Antennas in Wideband Head Imaging. *IEEE Antennas Wirel. Propag. Lett.* **2016**, *15*, 1618–1621. [CrossRef]
20. Hailu, D.M.; Safavi-Naeini, S. Narrow focus ultra-wideband antenna for breast cancer detection. In Proceedings of the IEEE Radio and Wireless Symposium, San Diego, CA, USA, 18–22 January 2009. [CrossRef]
21. Faisal, M.A.; Uddin, M.J.; Ullah, M.W.; Kamrul, M.I.; Haque, K.M.; Rahman, E. Comparative analysis of different types of breast cancer cell detection antennas. In Proceedings of the 2017 International Conference on Inventive Computing and Informatics (ICICI), Coimbatore, India, 23–24 November 2017. [CrossRef]

22. Sugitani, T.; Kubota, S.; Toya, A.; Kikkawa, T. Compact planar UWB antenna array for breast cancer detection. In Proceedings of the 2012 IEEE International Symposium on Antennas and Propagation, Chicago, IL, USA, 8–14 July 2012. [CrossRef]
23. Hasan, K.; Hadidy, M.E.; Morsi, H. Reflectarray antenna for breast cancer detection and biomedical applications. In Proceedings of the IEEE Middle East Conference on Antennas and Propagation (MECAP), Beirut, Lebanon, 20–22 September 2016. [CrossRef]
24. Iliopoulos, I.; Meo, S.D.; Pasian, M.; Zhadobov, M.; Pouliguen, P.; Potier, P.; Perregrini, L.; Sauleau, R.; Ettorre, M. Enhancement of penetration of millimeter waves by field focusing applied to breast cancer detection. *IEEE Trans. Biomed. Eng.* **2021**, *68*, 959–966. [CrossRef]
25. Ley, S.; Sachs, J.; Faenger, B.; Hilger, I.; Helbig, M. MNP-enhanced microwave medical imaging by means of Pseudo-Noise Sensing. *Sensors* **2021**, *21*, 6613. [CrossRef]
26. Botterill, T.; Lotz, T.; Kashif, A.; Chase, G. Reconstructing 3-D Skin Surface Motion for the DIET Breast Cancer Screening System. *IEEE Trans. Med. Imaging* **2014**, *33*, 1109–1118. [CrossRef]
27. Li, X.; Hagness, S.C. A confocal microwave imaging algorithm for Breast Cancer Detection. *IEEE Microw. Wirel. Compon. Lett.* **2001**, *11*, 130–132. [CrossRef]
28. Bond, E.J.; Li, X.; Hagness, S.C.; Van Veen, B.D. Microwave Imaging via space-time beamforming for early detection of breast cancer. In Proceedings of the IEEE International Conference on Acoustics Speech and Signal Processing, Orlando, FL, USA, 13–17 May 2002. [CrossRef]
29. Klemm, M.; Craddock, I.J.; Leendertz, J.A.; Preece, A.; Benjamin, R. Improved delay-and-sum beamforming algorithm for Breast Cancer Detection. *Int. J. Antennas Propag.* **2008**, *2008*, 761402. [CrossRef]
30. Özmen, H.; Kurt, M.B. Radar-based Microwave Breast Cancer Detection System with a high-performance ultrawide band Antipodal Vivaldi Antenna. *Turk. J. Electr. Eng. Comput. Sci.* **2021**, *29*, 2326–2345. [CrossRef]
31. Guruswamy, S.; Chinniah, R.; Thangavelu, K. Design and implementation of compact ultra-wideband Vivaldi antenna with directors for microwave-based imaging of breast cancer. *Analog. Integr. Circuits Signal Processing* **2021**, *108*, 45–57. [CrossRef]
32. Danjuma, I.M.; Noras, J.M.; Abd-Alhameed, R.A.; Obeidat, H.A.; Oguntala, G.A.; Eya, N.N.; Mohammad, B.A. Microwave imaging using arrays of Vivaldi antenna for breast cancer applications. *Int. J. Microw. Appl.* **2018**, *7*, 32–37. [CrossRef]
33. Islam, M.T.; Mahmud, M.Z.; Islam, M.T.; Kibria, S.; Samsuzzaman, M. A low cost and portable microwave imaging system for breast tumor detection using UWB directional antenna array. *Sci. Rep.* **2019**, *9*, 15491. [CrossRef] [PubMed]
34. O'Loughlin, D.; Elahi, M.A.; Porter, E.; Shahzad, A.; Oliveira, B.L.; Glavin, M.; Jones, E.; O'Halloran, M. Open-source software for microwave radar-based Image Reconstruction. In Proceedings of the 12th European Conference on Antennas and Propagation, London, UK, 9–13 April 2018. [CrossRef]

applied sciences

MDPI

Article

A Jug-Shaped CPW-Fed Ultra-Wideband Printed Monopole Antenna for Wireless Communications Networks

Sarosh Ahmad [1,2,*], Umer Ijaz [2], Salman Naseer [3], Adnan Ghaffar [4], Muhammad Awais Qasim [2], Faisal Abrar [2], Naser Ojaroudi Parchin [5,*], Chan Hwang See [5] and Raed Abd-Alhameed [6]

1 Department of Signal Theory and Communications, Universidad Carlos III de Madrid, Leganes, 28911 Madrid, Spain
2 Department of Electrical Engineering and Technology, Government College University Faisalabad (GCUF), Faisalabad 38000, Pakistan; umer.ijaz@gcuf.edu.pk (U.I.); imawais16@gmail.com (M.A.Q.); imfaisalabrar@gmail.com (F.A.)
3 Department of Information Technology, University of the Punjab Gujranwala Campus, Gujranwala 52250, Pakistan; salman@pugc.edu.pk
4 Department of Electrical and Electronic Engineering, Auckland University of Technology, Auckland 1010, New Zealand; aghaffar@aut.ac.nz
5 School of Engineering and the Built Environment, Edinburgh Napier University, Edinburgh EH14 1DJ, UK; C.See@napier.ac.uk
6 Faculty of Engineering and Informatics, University of Bradford, Bradford BD7 1DP, UK; r.a.a.abd@bradford.ac.uk
* Correspondence: saroshahmad@ieee.org (S.A.); N.OjaroudiParchin@napier.ac.uk (N.O.P.)

Abstract: A type of telecommunication technology called an ultra-wideband (UWB) is used to provide a typical solution for short-range wireless communication due to large bandwidth and low power consumption in transmission and reception. Printed monopole antennas are considered as a preferred platform for implementing this technology because of its alluring characteristics such as light weight, low cost, ease of fabrication, integration capability with other systems, etc. Therefore, a compact-sized ultra-wideband (UWB) printed monopole antenna with improved gain and efficiency is presented in this article. Computer simulation technology microwave studio (CSTMWS) software is used to build and analyze the proposed antenna design technique. This broadband printed monopole antenna contains a jug-shaped radiator fed by a coplanar waveguide (CPW) technique. The designed UWB antenna is fabricated on a low-cost FR-4 substrate with relative permittivity of 4.3, loss tangent of 0.025, and a standard height of 1.6 mm, sized at 25 mm × 22 mm × 1.6 mm, suitable for wireless communication system. The designed UWB antenna works with maximum gain (peak gain of 4.1 dB) across the whole UWB spectrum of 3–11 GHz. The results are simulated, measured, and debated in detail. Different parametric studies based on numerical simulations are involved to arrive at the optimal design through monitoring the effects of adding cuts on the performance of the proposed antennas. Therefore, these parametric studies are optimized to achieve maximum antenna bandwidth with relatively best gain. The proposed patch antenna shape is like a jug with a handle that offers greater bandwidth, good gain, higher efficiency, and compact size.

Keywords: printed monopole; CPW-fed; UWB; wireless communication

Citation: Ahmad, S.; Ijaz, U.; Naseer, S.; Ghaffar, A.; Qasim, M.A.; Abrar, F.; Parchin, N.O.; See, C.H.; Abd-Alhameed, R. A Jug-Shaped CPW-Fed Ultra-Wideband Printed Monopole Antenna for Wireless Communications Networks. *Appl. Sci.* **2022**, *12*, 821. https://doi.org/10.3390/app12020821

Academic Editor: Amalia Miliou

Received: 8 December 2021
Accepted: 12 January 2022
Published: 14 January 2022

1. Introduction

An ultra-wideband (UWB) is a telecommunications technology that is utilized in radio communication networks to achieve high-speed bandwidth connections with minimal energy consumption. Primarily, the UWB was intended for commercial radar. Wireless personal area networks (WPANs) and consumer electronics are two main applications of UWB technology. UWB wireless has developed as an emerging skill with limited smart structures such as radar, wireless communications, and medical engineering domains since its initial achievement in the middle of the 2000s [1]. Until 2001, UWB was significantly used for military purposes. The Federal Communications Commission (FCC) permitted the public to use UWB bandwidth for commercial purposes after 2002. Furthermore, the FCC approved the usage of the UWB spectrum, which is allocated between 3.1–10.6 GHz in the United States [2]. The low spectral density of UWB is responsible for short range of communication. This function, however, demands high-gain antennas with relatively stable radiation characteristics [3]. Planar antennas, primarily monopoles, are used in UWB electrical devices [4,5] due to its compact size, low profile, and low cost, as well as its ultra-wide impedance bandwidth. Moreover, when these antennas are placed near metallic surfaces, they can cause severe impedance mismatch. Low-profile antennas also transmit limited frequency signals with low gain and poor directivity [6,7].

However, the cost and size of the UWB antennas increases with discreet filters [8]. Frequencies from 5.2–5.8 GHz were notched by etching an omega type slot on the surface of the antenna in [9]. Similarly, in [10], U- and inverted U-shaped slots were embedded in printed monopoles to stop multiple frequencies. A curved shaped slot is introduced in [11] to achieve notching features in WiMAX and WLAN bands. To attain notching characteristics in 5.10–5.94 GHz, an S-shaped slot is applied in the feedline of the monopole antenna in [12]. Split ring resonators can act both as a band stop and band pass filters for different frequencies [13]. In [14], uplink and downlink satellite frequency bands were rejected by introducing a single SRR slot in the patch of the antenna. A WLAN band is rejected by inserting split ring resonators in [15]. Three different frequencies were notched in [16] by embedding multiple split rings near the feedline of the antenna model. Notching was achieved by using SRR in [17]. Tri-notching using frequency-selective surface (FSS) of an ultra-wideband antenna with gain augmentation was reported in [18]. Another compact-sized UWB planar antenna using truncated ground plane was presented in [19]. The antenna covers large bandwidth, but the size was still large, compared to our design. A broadband overleaf-shaped antenna using beam tilt characteristics is presented in [20]. The reported size of the antenna is large, and small bandwidth is achieved, as shown in Table 1. Another Vivaldi antenna resonative over a wide frequency range is reported in [21]. The antenna is antipodal, and the miniaturization was achieved by using exponential strip arms technique.

Table 1. Comparison with the previous research.

Ref. No.	Frequency Range (GHz)	Area (mm^2)	Electrical Size ($\lambda_0{}^2$)	Antenna Type	Substrate Material	Efficiency (%)	Gain (dB)
[11]	3.4–7, 8–11.4	40 × 30	0.94 × 0.705	Split Ring Resonator Patch	FR-4	<95	<5
[12]	3.1–10.6	38.31 × 34.52	0.82 × 0.74	Monopole	FR-4	<95	<5
[13]	4.05–5.1, 6–13	32 × 36	0.89 × 1.01	Circular Patch	FR-4	-	<4
[14]	2.5–19.8	36 × 25	0.62 × 0.43	Slotted Patch	FR-4	-	<3
[15]	2.8–18	50 × 38	0.96 × 0.73	Tapered Slotted Patch	FR-4	-	<4.32
[16]	1.9–5, 6–10.6	48 × 55	0.63 × 0.72	Monopole	FR-4	-	<5
[17]	1.2–9.8	53 × 63.5	0.21 × 0.25	Anti-Spiral Shaped Patch	FR-4	<85	<5.2
[18]	2.6–10.58	38.3 × 34.5	0.33 × 0.3	Sharp triple notched	FR-4	-	<5
[19]	1.5–10.4	64 × 37.4	0.32 × 0.19	Planar patch	F4BM	-	>2
[20]	2–5	100 × 78	0.67 × 0.52	Leaf Shaped Patch	Taconic TLY-5	-	>3
[21]	0.83–9.8	161 × 140	0.45 × 0.39	Ex-potential Strip Arms	$\varepsilon_r = 2.3$	-	>2.5
This Work	**3–11**	**25 × 22**	**0.25 × 0.22**	**Printed Monopole**	**FR-4**	**>85**	**<4.1**

In this research article, a simple CPW based an ultra-wideband antenna having impedance bandwidth ranging from 3 GHz to 11 GHz (8 GHz) for wireless communication networks is presented. It is very hard to achieve a UWB band with compact size; however, in this design, the UWB band is achieved through a CPW technique and the design optimization. The total size of the designed UWB antenna is 25 mm × 22 mm × 1.6 mm. This printed broadband monopole antenna is manufactured using a low-cost FR-4 Duroid material. The antenna presents good efficiency with suitable gain. This article is organized as follows: the proposed antenna design is presented in Section 2. Results and discussion are presented in Section 3, and the conclusion is given in Section 4.

2. Antenna Design Analysis

The schematic diagram of the designed ultrawideband antenna is presented in Figure 1. The structure of the UWB antenna involves the jug-shaped printed monopole with handle at the right side of the radiator, sized at Ls × Ws × hs. The printed monopole is fed by a coplanar waveguide (CPW) feedline of length "Lf" and width "wf". The width of the CPW feedline is kept at 3 mm to attain 50 Ω input impedance. The antenna is designed on a low-cost FR-4 substrate having relative permittivity (ε_r) of 4.4 and loss tangent (tanδ) of 0.025. The design is simulated in computer simulation technology (CST-2018) software. On the front view of the substrate, a rectangular ground plane is designed having dimension Lg × Wg, and the ground plane is shown. The three-dimensional (3D) view of the antenna is depicted in Figure 1, and its optimized dimensions are given in Table 2.

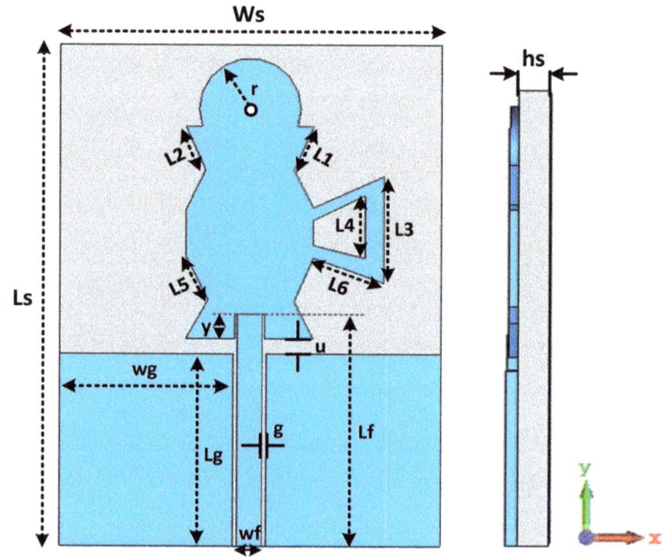

Figure 1. Schematic diagram of the presented UWB antenna.

Table 2. Different design parameters of the presented antenna.

Parameters	Values (mm)	Parameters	Values (mm)
Ls	25	Ws	22
Lf	14.65	Wf	1.58
Lg	12.14	Wg	10.85
L1	3.02	L2	2.75
L3	6.73	L4	3.96
L5	3.05	L6	4.64
g	0.24	y	1.58
u	0.95	hs	1.6

2.1. Different Design Steps

Figure 2 shows the different design steps of the designed monopole and the S_{11} behavior given in Figure 3. In the first step, the basic design consists of a simple rectangular printed monopole radiator excited by a coplanar waveguide (CPW) feedline, as shown in ANT I. Then, in the second step, the simple rectangular radiator is truncated from its upper and lower sides to keep its shape similar to the body of a jug, which helps to keep the S_{11} (dB) close to −10 dB, but the antenna only operates at 3.5 GHz and 10.5 GHz. Again, in the third step, a semicircular-shaped patch is introduced in the ANT II, which keeps some portion of the UWB band below −10 dB, but the antenna works from 3.3 GHz to 9 GHz and 9.3 GHz to 12 GHz, as can be seen in ANT III (Figure 3), and this is not a required frequency band. Now, in order to achieve the whole UWB spectrum from 3 GHz to 11 GHz, a C-shaped resonator is introduced in the final step to make the shape similar to a handle of a jug, as shown in ANT IV.

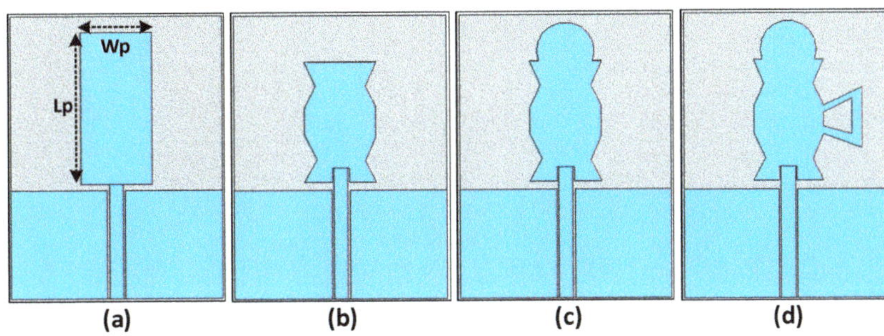

Figure 2. Design steps of the presented ultra-wideband antenna: (**a**) rectangular printed monopole only (ANT I), (**b**) truncated monopole (ANT II), (**c**) addition of semicircular printed monopole (ANT III), (**d**) presented design (ANT IV).

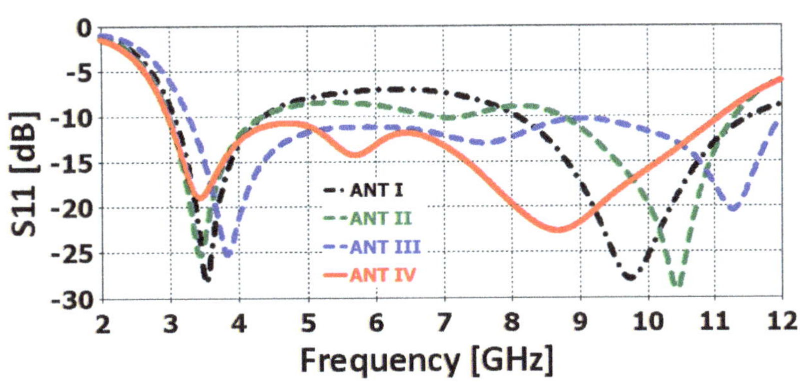

Figure 3. S_{11} (dB) of the different design steps of the presented UWB antenna shown in Figure 2.

The design process of the printed monopole antenna is explained as follows: The primary antenna design (ANT I), shown in Figure 3, contains a 50 Ω CPW feedline, a jug-shaped monopole, and the ground plane. The printed monopole's width and length are calculated using Equations (1) and (2) [13], as follows:

$$Wp = \frac{\lambda_o}{2(\sqrt{0.5(\varepsilon_r + 1)}} \tag{1}$$

where ε_r and λ_o are the relative permittivity and the wavelength of the substrate in free space at the operating frequency. The best option of *Wp* tends to the perfect impedance matching. The length of the printed monopole can be evaluated by using Equation (2).

$$Lp = \frac{c_o}{2f_o\sqrt{\varepsilon_{eff}}} - 2\Delta L_p \quad (2)$$

where c_o, ΔL_p, and ε_{eff} are the velocity of light, change in the length of the printed monopole due to its fringing effect, and the effective dielectric constant, respectively. The effective relative permittivity can be calculated using Equation (3):

$$\varepsilon_{eff} = \frac{\varepsilon_r + 1}{2} + \frac{\varepsilon_r - 1}{2}\left(\frac{1}{\sqrt{1 + 12\frac{hs}{W_p}}}\right) \quad (3)$$

where *hsub* is the height of the substrate. At the end, the fringing effect can be calculated using Equation (4):

$$\Delta Lp = 0.421 h_s \frac{(\varepsilon_{eff} + 0.300)(\frac{W_p}{hs} + 0.264)}{(\varepsilon_{eff} - 0.258)(\frac{W_p}{hs} + 0.813)} \quad (4)$$

With the placement of $\varepsilon_r = 4.3$, *hs* = 1.6 mm in Equations (1)–(4), the initial parameters of the rectangular printed monopole are *Lp* = 15 mm and *Wp* = 12 mm.

With a simple rectangular monopole (ANT I), the antenna works only works at 3.5 GHz and 9.8 GHz, as shown in Figure 3. By ANT II, the bandwidth of the antenna is increased but is unable to achieve UWB band. Then, in the third step (ANT III), with the help of the semicircular printed monopole on the top of the truncated printed monopole, the antenna achieves most of the UWB band, as the antenna has achieved band from 3 GHz to 11 GHz

2.2. Parametric Study of the Presented Design

The presented design is finalized after performing several parametric optimizations on different variables, as shown in Figure 4. The first parametric study is performed on the length and width of the ground plane. By increasing the length of the ground plane "Lg" from 10 mm to 12 mm, the impedance matching of the antenna improves with suitable bandwidth, and when the width of the ground plane "wg" is varied from 9 mm to 11 mm, then the bandwidth of the antenna is increased from 4.1 GHz to 8 GHz. The next parametric study is performed on the width of the feedline "wf". Gradually increasing the width of the feedline improves the impedance bandwidth from 5.8–8 GHz. A parametric study of the C-shaped radiator is also performed. By varying the lengths "L6" and "L3", the bandwidth of the antenna is improved, as depicted in Figure 4.

The surface current densities of the UWB antenna at different frequency bands are taken into consideration. This indicates that the antenna plays a significant role in making it resonate at the desired frequency bands. For example, the surface current density at 3.5 GHz is illustrated in Figure 5a. Most of the current seems to flow through the radiator at 3.5 GHz (see Figure 5b), while at 4.1 GHz, the current only flows through the C-shaped resonator and some amount of current through the feedline (see Figure 5c). At 8 GHz, the current flows through the outer lower edge of the printed monopole and some amount of current flow through the CPW ground at 10.5 GHz (see Figure 5d).

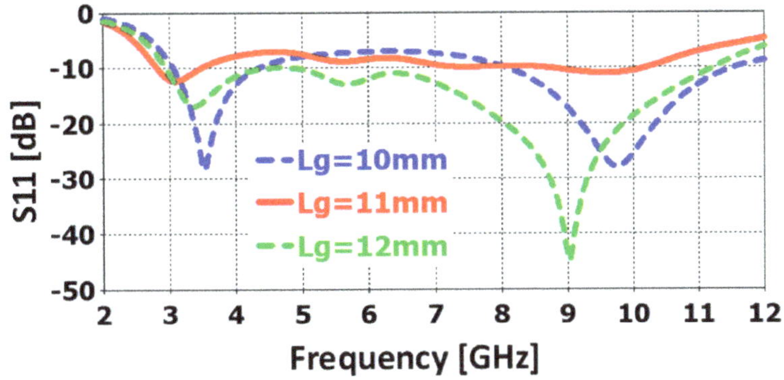

(a)

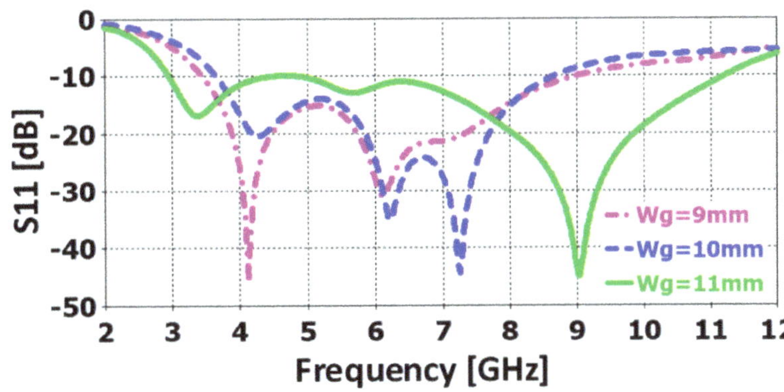

(b)

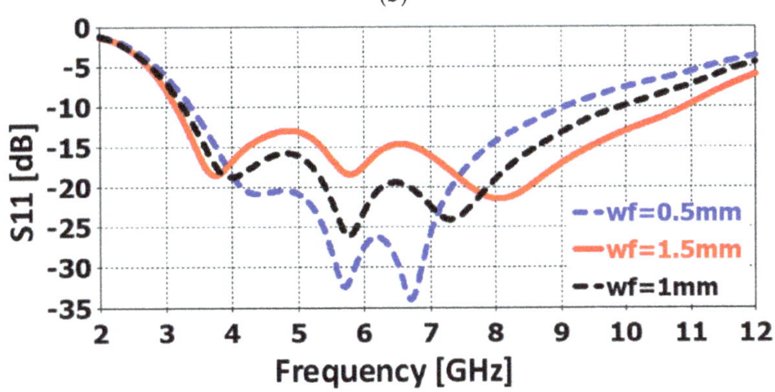

(c)

Figure 4. *Cont.*

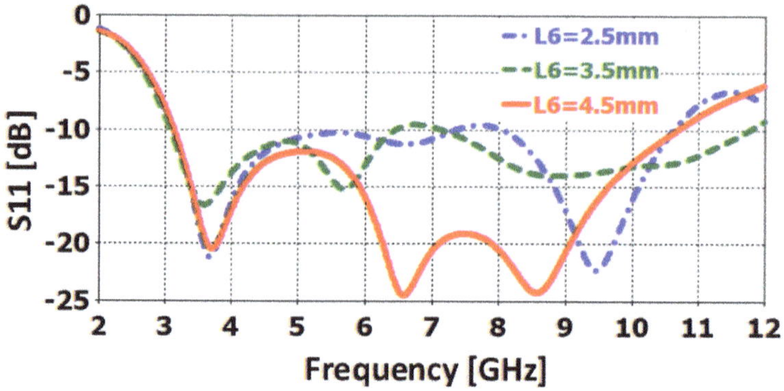

(d)

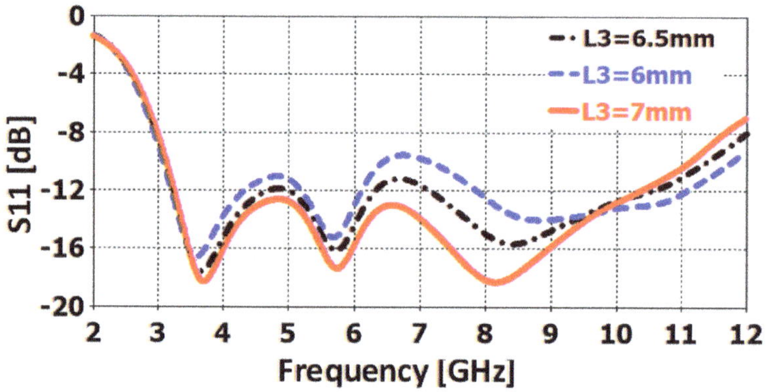

(e)

Figure 4. Different parameters optimization: (**a**) variation in "Lg", (**b**) variation in "wg", (**c**) variation in "wf", (**d**) variation in "L6", (**e**) variation in "L3".

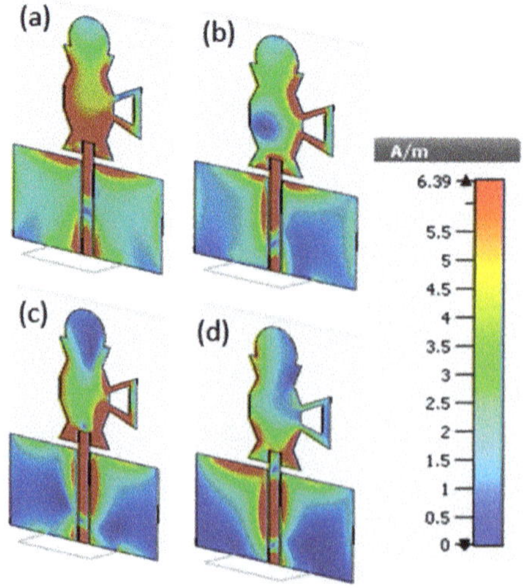

Figure 5. Surface current density (**a**) at 3.5 GHz, (**b**) at 4.1 GHz, (**c**) at 8 GHz, (**d**) at 10.5 GHz.

2.3. Equivalent Circuit Model

A circuit model for the UWB presented antenna for wireless communications is presented in Figure 6a. The main purpose of the circuit model is to validate the scattering parameters of the ultra-wideband antenna with the S_{11} obtained from the circuit model. The circuit model is designed by using advanced design system (ADS) software. The circuit model consists of four inductors, four capacitors, three resistors, and three resistor–capacitor (RC) circuits connected in series with one resistor and an inductor for each, as given in Figure 6a. By varying the values of the resistors, the S_{11} of the circuit model can be varied, while by fluctuating the values of the capacitors and inductors, the S_{11} of the antenna can be tuned. The values of the lumped components are given in Table 3. The S_{11} (dB) of the circuit model is illustrated in Figure 6b. It covers the bandwidth from 3.1 GHz to 11.5 GHz.

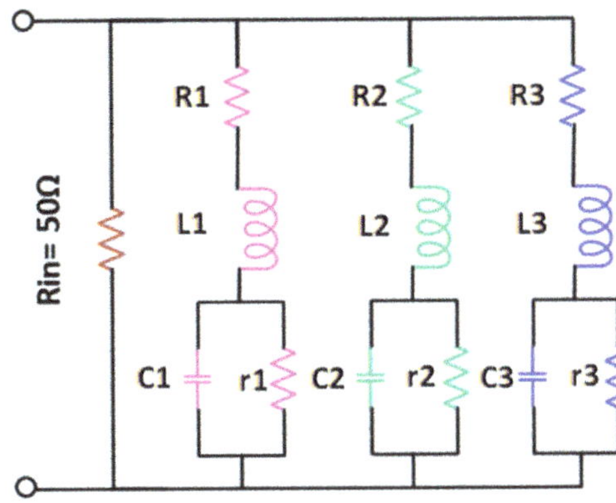

(**a**)

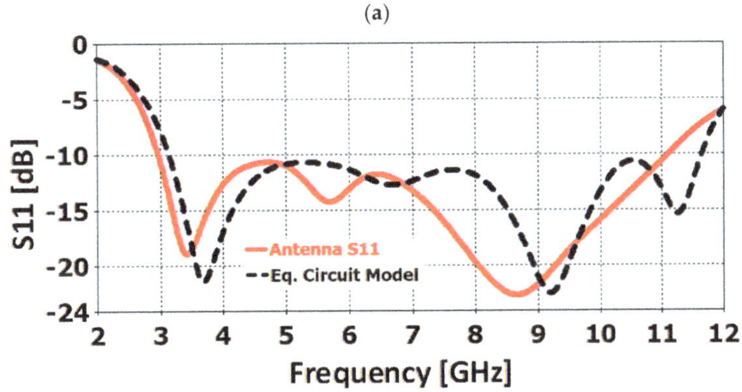

(**b**)

Figure 6. (**a**) Equivalent circuit model, and (**b**) reflection coefficient of the equivalent circuit model.

Table 3. Values of the components used in the circuit model.

Capacitors	Values (pF)	Inductors	Values (nH)	Resistors	Values (Ω)	High Resistors	Values (Ω)
C1	1	L1	7	R1	2	r1	1500
C2	0.1	L2	0.8	R2	65	r2	1000
C3	0.5	L3	0.5	R3	65	r3	500

3. Results and Discussions

In order to measure the scattering parameters of the fabricated prototype, the port of the fabricated design is connected with a vector network analyzer (VNA). The picture of the prototype is visible in Figure 7a. The S_{11} (dB) of the projected antenna is accessible in Figure 7b. Due to intolerances in the fabrication process and surrounding noises, there are some variations in the measured results. The simulated and measured S_{11} (dB) are in good agreement, as both are covering the whole UWB band for wireless communications.

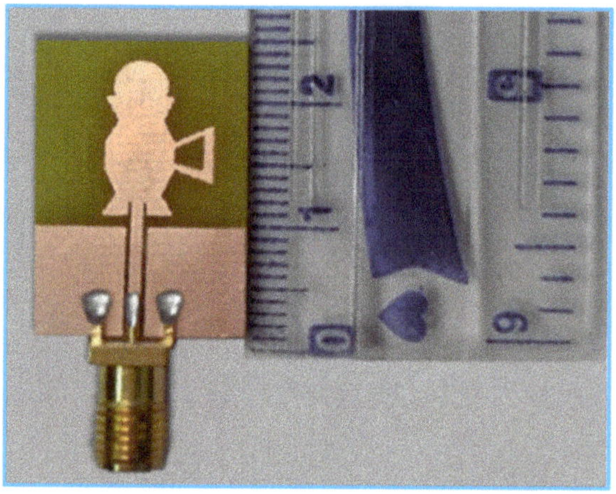

(a)

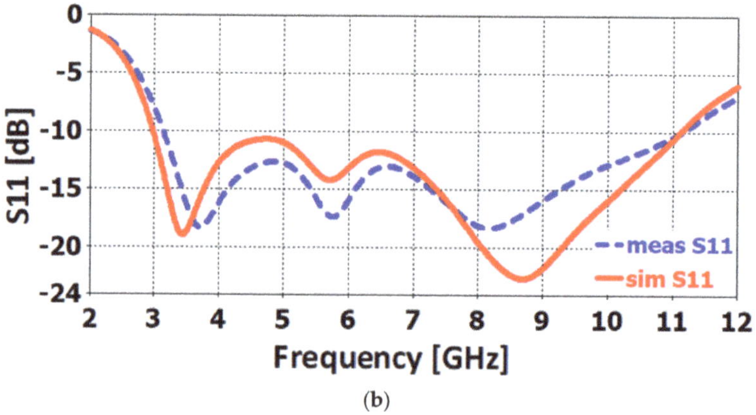

(b)

Figure 7. (**a**) Printed UWB prototype; (**b**) comparison of simulated and measured reflection coefficients (S_{11}).

The simulated and measured (E & H) plane of the UWB antenna at 3.5 GHz, 4.1 GHz, 8 GHz, and 10.5 GHz are given in Figure 8. It can be seen that there is an omnidirectional pattern at the frequencies of 3.5 GHz and 4.1 GHz along the E-plane while elliptical along the H-plane, and the antenna has a broadside radiation pattern in both planes at the frequencies of 8 GHz and 10.5 GHz. The simulated and measured gain graph is presented in Figure 9. It can be noticed that the antenna has attained the average peak gain ranges from 2–4.1 dB and the antenna's efficiency is attained for more than 85% over the entire band. A comparison with the previous research is given in Table 1.

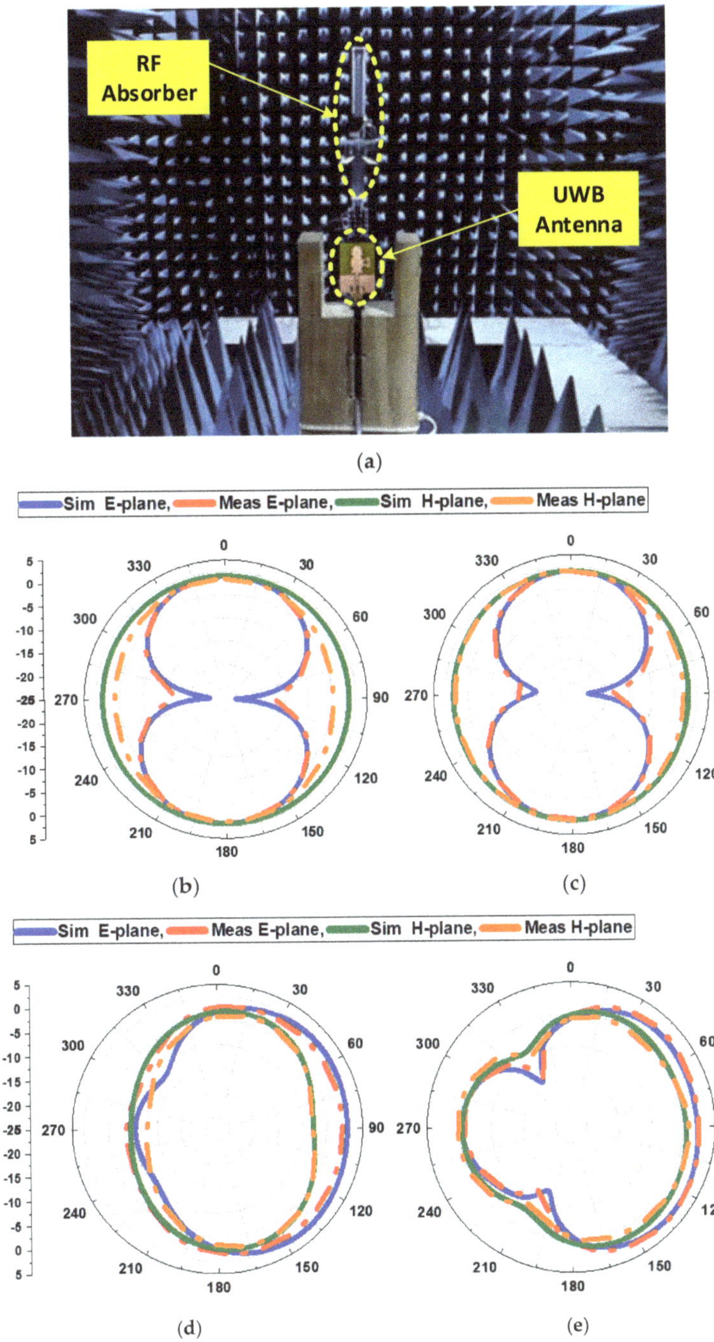

Figure 8. (**a**) Simulated and measured 2D radiation pattern setup inside chamber (**b**) at 3.5 GHz, (**c**) at 4.1 GHz, (**d**) at 8 GHz, (**e**) at 10.5 GHz.

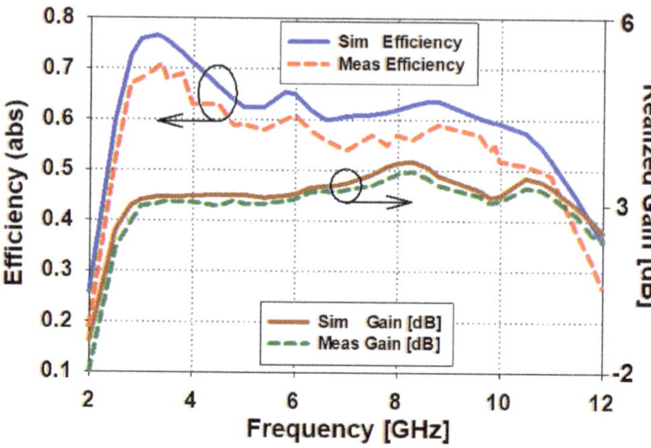

Figure 9. Comparison of simulated and measured efficiency and peak gains (dB).

4. Conclusions

A simple jug-shaped ultra-wideband (UWB) antenna is presented in this work. The presented design is printed, and measured results are also taken. The simulated results are verified by a measured result of the ultra-wideband antenna. The designed UWB antenna is printed on a low-cost FR-4 substrate with relative permittivity of 4.3, loss tangent 0.025, and a standard thickness 1.6 mm, sized at 25 mm × 22 mm × 1.6 mm, suitable for wireless communication system. The designed UWB antenna works with maximum gain (peak gain of 4.1 dB) across the whole UWB spectrum of 3–11 GHz. The simulated and measured reflection coefficients and radiation pattern are in close agreement. The designed antenna is a good applicant for wireless communication systems portable devices.

Author Contributions: Conceptualization, S.A., S.N., M.A.Q. and F.A.; methodology, U.I., A.G. and C.H.S.; software, A.G., S.N. and U.I.; validation, M.A.Q., F.A., U.I. and S.N.; formal analysis, S.A. and A.G.; investigation, A.G., N.O.P. and S.N.; resources, A.G., U.I. and S.A.; data curation, S.A., M.A.Q. and A.G.; writing—original draft preparation, S.A. and A.G.; writing—review and editing, S.N. and F.A.; supervision, N.O.P., C.H.S. and R.A.-A.; funding acquisition, N.O.P., C.H.S. and R.A.-A. All authors have read and agreed to the published version of the manuscript.

Funding: This work received no funding.

Conflicts of Interest: The authors declare no conflict of interest.

References

1. Kirtania, S.; Younes, B.; Hossain, A.; Karacolak, T.; Sekhar, P. CPW-Fed Flexible Ultra-Wideband Antenna for IoT Applications. *Micromachines* **2021**, *12*, 453. [CrossRef] [PubMed]
2. Amdaouch, I.; Aghzout, O.; Naghar, A.; Alejos, A.V.; Falcone, F. Design of UWB compact slotted monopole antenna for breast cancer detection. *Adv. Electromagn.* **2019**, *8*, 1–6. [CrossRef]
3. Saeidi, T.; Ismail, I.; Wen, W.P.; Alhawari, A.R.H.; Mohammadi, A. Ultra-Wideband antennas for wireless commu-nication applications. *Int. J. Antennas Propag.* **2019**, *19*, 7918765.
4. Alhawari, A.; Majeed, S.; Saeidi, T.; Mumtaz, S.; Alghamdi, H.; Hindi, A.; Almawgani, A.; Imran, M.; Abbasi, Q. Compact Elliptical UWB Antenna for Underwater Wireless Communications. *Micromachines* **2021**, *12*, 411. [CrossRef] [PubMed]
5. Babale, S.A.; Paracha, K.N.; Ahmad, S.; Abdul Rahim, S.K.; Yunusa, Z.; Nasir, M.; Ghaffar, A.; Lamkaddem, A. A Recent Approach towards Fluidic Microstrip Devices and Gas Sensors: A Review. *Electronics* **2022**, *11*, 229. [CrossRef]
6. Li, X.-P.; Xu, G.; Ma, M.-R.; Duan, C.-J. UWB Dual-Band-Notched Lanky-Leaf-Shaped Antenna with Loaded Half-Square-Like Slots for Communication System. *Electronics* **2021**, *10*, 1991. [CrossRef]
7. Rasool, J.M. Ultra Wide Band Antenna Design for Robotics & Sensors Environment. In Proceedings of the 2019 12th International Conference on Developments in eSystems Engineering (DeSE), Kazan, Russia, 7–10 October 2019; pp. 668–672.
8. Iqbal, A.; Smida, A.; Mallat, N.K.; Islam, M.T.; Kim, S. A Compact UWB Antenna with Independently Controllable Notch Bands. *Sensors* **2019**, *19*, 1411. [CrossRef] [PubMed]

9. Abbas, A.; Hussain, N.; Lee, J.; Park, S.G.; Kim, N. Triple Rectangular Notch UWB Antenna Using EBG and SRR. *IEEE Access* **2020**, *9*, 2508–2515. [CrossRef]
10. Abbas, A.; Hussain, N.; Jeong, M.-J.; Park, J.; Shin, K.S.; Kim, T.; Kim, N. A Rectangular Notch-Band UWB Antenna with Controllable Notched Bandwidth and Centre Frequency. *Sensors* **2020**, *20*, 777. [CrossRef] [PubMed]
11. Yeboah-Akowuah, B.; Tchao, E.T.; Ur-Rehman, M.; Khan, M.M.; Ahmad, S. Study of a printed split-ring monopole for dual-spectrum communications. *Heliyon* **2021**, *7*, e07928. [CrossRef] [PubMed]
12. Alam, M.S.; Abbosh, A. Reconfigurable band-rejection antenna for ultra-wideband applications. *Microw. Antennas Propag. IET* **2018**, *12*, 195–202. [CrossRef]
13. Abdi, H.; Nourinia, J.; Ghobadi, C. Compact Enhanced CPW-Fed Antenna for UWB Applications. *Adv. Electromagn.* **2021**, *10*, 15–20. [CrossRef]
14. Upadhyay, A.; Khanna, R. A CPW-fed tomb shaped antenna for UWB applications. *Int. J. Innov. Technol. Explor. Eng. (IJITEE)* **2019**, *8*, 67–72.
15. Hassain, Z.A.A.; Azeez, A.R.; Ali, M.M.; Elwi, T.A. A modified compact bi-directional UWB tapered slot antenna with double band notch characteristics. *Adv. Electromagn.* **2019**, *8*, 74–79. [CrossRef]
16. Chaudhary, P.; Kumar, A. Compact ultra-wideband circularly polarized CPW-fed monopole antenna. *AEU—Int. J. Electron. Commun.* **2019**, *107*, 137–145. [CrossRef]
17. Li, X.-P.; Xu, G.; Duan, C.-J.; Ma, M.-R.; Shi, S.-E.; Li, W. Compact TSA with Anti-Spiral Shape and Lumped Resistors for UWB Applications. *Micromachines* **2021**, *12*, 1029. [CrossRef] [PubMed]
18. Kundu, S.; Chatterjee, A. Sharp Triple-notched ultra-wideband antenna with gain augmentation using FSS for ground pen-etrating radar. *Wirel. Pers. Commun.* **2021**, *117*, 1399–1418. [CrossRef]
19. Guo, L.; Min, M.; Che, W.; Yang, W. A Novel Miniaturized Planar Ultra-Wideband Antenna. *IEEE Access* **2018**, *7*, 2769–2773. [CrossRef]
20. Delphine, A.; Hamid, M.R.; Seman, N.; Himdi, M. Broadband cloverleaf Vivaldi antenna with beam tilt characteristics. *Int. J. RF Microw. Comput. Eng.* **2020**, *30*, e22158. [CrossRef]
21. Honari, M.M.; Ghaffarian, M.S.; Mirzavand, R. Miniaturized Antipodal Vivaldi Antenna with Improved Bandwidth Using Exponential Strip Arms. *Electronics* **2021**, *10*, 83. [CrossRef]

applied sciences

MDPI

Article

Performance Enhancement of Reconfigurable Metamaterial Reflector Antenna by Decreasing the Absorption of the Reflected Beam

Efi Rahamim *, David Rotshild and Amir Abramovich

Department of Electrical and Electronic Engineering, Ariel University, Ariel 407000, Israel; drotshild@gmail.com (D.R.); amir007@ariel.ac.il (A.A.)
* Correspondence: efirahamim1@gmail.com

Abstract: In this study, a new concept for a Ka-band 5G communication tunable reflector metasurface (MS) for beam steering at 28 GHz is proposed. Varactor diodes are used as the tunability component of each unit cell of this MS. Significant improvements in beam steering and bandwidth performance were achieved using this new concept referred to as the stripes configuration. Several different geometries of unit cells arranged in stripes were designed to achieve better performance in directionality, gain, sidelobe level (SLL), and bandwidth in the stripes configuration. Simulation results for a three-stripe MS with different unit cells in each stripe showed better performance in the phase dynamic range and reduced reflectance losses compared to a typical one-stripe MS. The simulation results showed a significant improvement of 3 dB, depending on the steering angle in reflectance gain, compared to a uniform MS (one stripe). Furthermore, a significant improvement of approximately 50% in the accuracy of the steering angle for different operating frequencies was demonstrated. Manufacturing considerations are discussed in this study.

Keywords: beam steering; Ka-band; metasurface; varactor diode

Citation: Rahamim, E.; Rotshild, D.; Abramovich, A. Performance Enhancement of Reconfigurable Metamaterial Reflector Antenna by Decreasing the Absorption of the Reflected Beam. *Appl. Sci.* **2021**, *11*, 8999. https://doi.org/10.3390/app11198999

Academic Editor: Naser Ojaroudi Parchin

Received: 18 July 2021
Accepted: 15 September 2021
Published: 27 September 2021

1. Introduction

A metasurface (MS) is a 2D metamaterial that has electromagnetic properties that do not exist in nature. In the last two decades, ag reat deal of research has been conducted in the field of metamaterials and MSs, enabling fascinating electromagnetic properties that cannot be achieved with naturally occurring materials. In contrast to conventional materials, MSs provide a new way to manipulate microwave and MMW radiation based on reflection from sub-wavelength unit cell periodic arrays, which simultaneously have negative permeability and permittivity [1–3]. The unique interaction between the unit cells and the electromagnetic radiation aids in the design of innovative components. Applications such as metasurface perfect absorbers [4], cloaks [5], flat parabolic mirror surfaces (FLAPSs) [6], and polarization control [7,8] have been demonstrated experimentally.

In 1968, Veselago [9] theoretically investigated the electrodynamic consequences of a medium having negative values of both ε and μ, and concluded that such a medium would have dramatically different propagation characteristics stemming from the sign change of the group velocity. However, these effects could not be experimentally verified since substances with $\mu < 0$ were not available. The split ring resonator (SRR) medium proposed by Pendry et al. [10] provided the opportunity to make a material with negative permeability, from which a "left-handed" medium could be constructed. By combining the SRR with cut wire in an array structure, a left-handed material [11] was created.

Reconfigurable metasurfaces (MSs) have attracted a great deal of interest in the last two decades. There are many interdisciplinary applications that have been realized by MSs [12]. Applications such as a millimeter-wave sensitive sensor MS based on a perfect absorber for the detection and recognition of micro-poisons in drinking water [13] and a W-band flat parabolic mirror for imaging systems and beam propagation [6] are resonant, and thus they are limited to a narrow bandwidth around the resonant frequency. Adding tunability to each unit cell significantly expands the performance and versatility of the MS for innovative applications. Reconfigurability of the MS can be realized by PIN switches [14], MEMS switches [15], liquid crystal or piezoelectric materials [16,17], and varactor diodes [8]. The PIN switch enables MSs with a fast switching time, high reliability, low cost, and low voltage control [15]. Tunable components such as MEMS switches have attracted attention due to their small size and linearity; on the other hand, they have a relatively slow switching time compared to PIN switch diodes [16]. The discrete nature of switches cannot be used in applications where a continuous phase change is required. Liquid crystal (LC) changes the dielectric coefficient of the unit cell substrate by tens of percentage points as a function of applied DC bias. LC can be used at K-band frequencies and above [15]. LC has a slow response time compared to switches and varactors [18] and a small dynamic range relative to the varactor. A piezoelectric crystal allows limited flexibility proportional to the change in its physical size and requires a relatively high DC bias [17]. The varactor diodes enable a continuous and high dynamic range of capacitance, constant gamma for linear tuning, a convenient PCB assembly, and very low power consumption [19].

However, the solutions offered in the literature suffer from a bandwidth problem. This makes the task of finding an efficient and practical solution more difficult. Previous works, extensively based on CST simulation code, were implemented and experimentally tested for reconfigurable MS reflectors at Ku-band [20] and K-band [7]. The experimental results of these MSs are in excellent agreement with the simulation results for the reflection property values and beam-steering abilities. Based on those experimental results, we propose a new configuration concept named the stripes method. The stripes configuration showed a better performance in gain, SLL, and bandwidth.

2. Unit Cell Design

In this study, a combination of two types of unit cell geometries arranged in the same MS reflector were designed. Figure 1a–d shows a typical unit cell design consisting of a front side, back side, profile, and 3D view of two adjacent unit cells. This unit cell is composed of two PCB layers made by the Rogers Company, model RT/Duroid 5880 with εr = 2.2 [21], and three copper layers of 0.5 oz thickness. The unit cell is composed of two vertical stripes in the top layer with each electrical connected to a pad, in conjunction with a varactor diode (used in this study) by a horizontal appropriate microstrip line. The varactor diode, model MAVR–011020–1411 of Macom [19], is used in this MS design due to its extremely low capacitance. The varactor diode is placed on the pad between the strips (see Figure 1 in green), adding variable capacitance $Cdiode$ to the unit cell. The dynamic range of the capacitance is $Cdiode\ max$ = 0.216 pF to $Cdiode\ min$ = 0.032 pF for 0–15 V reverse bias voltage, respectively. One of the vertical stripes has the same length as the unit cell width, W (used for the DC ground), while the second strip, S_L, has a shorter length by 0.05 mm and is connected through a via to the back side of the PCB for the DC bias circuit. The middle copper layer acts as a metal ground for the unit cell, and it is separated by clearance (CD) from the via, as shown in Figure 1d. To have a symmetrical unit cell design, a mirror unit cell was added, as seen in Figure 1, where polarization- and propagation-instated reduction hits are perpendicular to the surface, as shown in Figure 1a.

In this study, we propose a new concept referred to as the stripes configuration. The proposed configuration is composed of three stripes (1, 2, and 3) of two types of MS reflectors (Type A and type B) arranged side by side, as can be seen in Figure 2. Each type has a different unit cell design as shown in Table 1. The designed MS has three stripes with each stripe composed of four unit cell columns (Figure 2). Each stripe has a different dielectric substrate thickness, h, yielding a smaller dynamic phase range improving the reflected gain. Stripe one and stripe three are identical stripes with a thickness of h = 0.6 mm. The thickness of stripe two, in the middle, is h = 0.75 mm. Using the 3D electromagnetic simulation software CST, three different unit cell parameters were investigated and optimized for best performance. The final dimensions are summarized in Table 1. Two different MSs for 28 GHz were designed and investigated, based on the unit cells given in Table 1. The first is a typical uniform MS (one stripe) based on a uniform unit cell. The second MS is composed of two types of unit cells, type A and type B, arranged in three stripes, as shown in Figure 2.

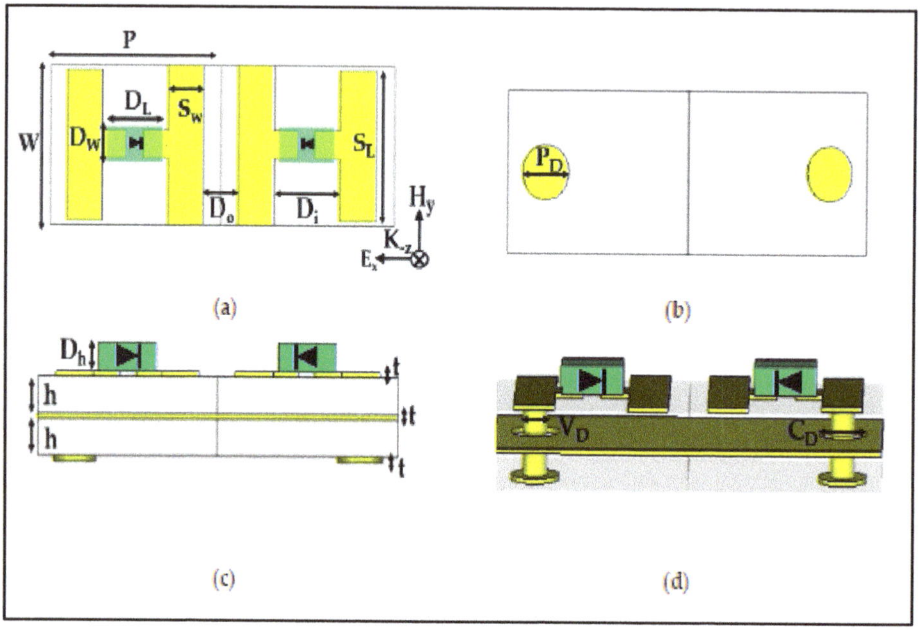

Figure 1. The design of two adjacent unit cells. (**a**) Unit cell top layer. (**b**) Unit cell back layer. (**c**) Side view of the unit cell. (**d**) 3D view of the unit cell [8].

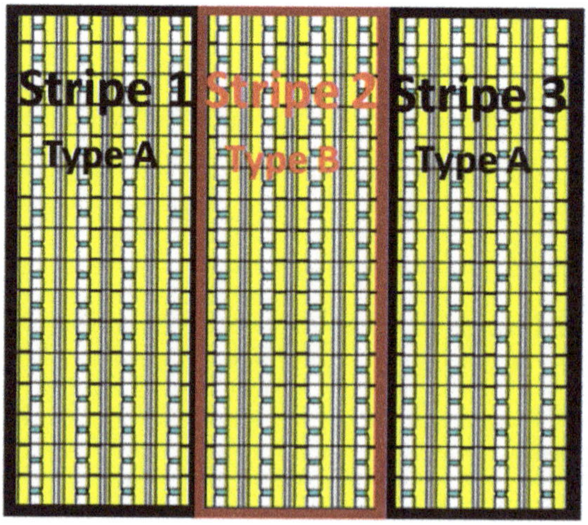

Figure 2. MS structure with three stripes stuck together.

There are three absorption-loss mechanisms in the proposed tunable MS: varactor diode losses caused by internal resistance of the diode (*Rdiode*), dielectric losses, and ohmic losses [8]. The dielectric losses and ohmic losses are well defined and quantified in the CST simulation. On the other hand, the varactor serial resistance *Rdiode* is unknown, therefore experimental measurements from similar studies were used [22,23], taking manufacturing process imperfections into consideration. The approximate value of *Rdiode* was found to be 4 Ω [22,23]. Table 2 summarizes the electrical parameters that were used in the simulations.

Table 1. Geometric unit cell parameters.

Parameter	Description	Value (mm) Uniform Unit Cell	Value (mm) Type A	Value (mm) Type B
P	Unit cell length	3.4	3.4	3.4
W	Unit cell width	2.55	2.55	2.55
S_L	Pad/Line(strip) length	2.45	2.45	2.45
S_w	Pad/Line(strip) width	0.9	0.9	0.9
D_L	Varactor diode length	0.7615	0.7615	0.7615
D_w	Varactor diode width	0.406	0.406	0.406
D_H	Varactor diode height	0.203	0.203	0.203
D_i	Distance between external line	0.686	0.686	0.686
D_o	Distance between internal line	0.94	0.94	0.94
h	Dielectric substrate thickness	0.381	0.6	0.75
t	Copper thickness	0.035	0.035	0.035
Pd	Pad diameter	0.6	0.6	0.6
VD	Via diameter	0.3	0.3	0.3
CD	Clearness diameter	0.6	0.6	0.6

Table 2. Electrical unit cell parameters.

Parameter	Description	Value
Rdiode	Diode resistance	4 Ω
Cdiode max	Max diode capacitance	0.216 pF
Cdiode min	Min diode capacitance	0.032 pF

The proposed MS reflector was simulated using the TEM plane wave port with CST simulation software [24]. The simulation results of the uniform unit cell, including magnitude and phase reflection as a function of frequency, are shown in Figure 3. The simulations were repeated for three capacitance values: *Cdiode min* = 0.032 pF (red), *Cdiode max* = 0.216 pF (green), and *Cdiode res* (28 GHz). This provided a dynamic reflection phase range of 311°, marked by a continuous black crosser, as shown in Figure 3a. The magnitude of the reflected beam for capacitance is shown in Figure 3b. The maximum loss value for *Cdiode res* (28 GHz) = 0.0815 pF was 5.5 dB, marked by a continuous black crosser, as shown in Figure 3b. The via was used for the DC bias for the varactor diode. The standard via diameter for manufacturing is 0.25 mm–0.30 mm, which had no effect on the MS's performance. Figure 3c describes the influence of the via diameter on the resonance frequency and the absorption, where no significant influence was found.

The simulation results of type A and type B unit cell phase reflection and magnitude are shown in Figure 4 as a function of frequency for three capacitance values: *Cdiode min* = 0.032 pF (green), *Cdiode max* = 0.216 (red) pF, and the 28 GHz resonance capacitance *Cdiode res*. Type A provided a dynamic reflection phase range of 236° and type B provided a dynamic reflection phase range of 150°, marked by the dashed black crossers shown in Figure 4a,c, respectively. The maximum loss value of type A *Cdiode res* (28 GHz) = 0.05 pF was 1.6 dB, and the maximum loss value of type B *Cdiode res* (28 GHz) = 0.04 pF was 0.8 dB, marked by continuous black crossers in Figure 4b,d, respectively.

The resonance frequency is related to the inductance and capacitance of the unit cell. The thickness, h, of the dielectric substrate is proportional to the inductance [25]; thus, the change of h affects the resonance frequency and the dynamic phase range of the unit cell. The motivation to design an MS based on the stripes method was to reduce the total absorption of the proposed MS compared to the uniform type (one stripe). Changing h is advantageous since it reduces the "transition effects" (or "edge effects"), yielding a better beam-steering performance. Any differences between two adjacent stripes would disturb the periodicity of the unit cells of the whole MS, causing transition effects or edge effects, degrading the reflected beam. The advantage of changing h (the dielectric substrate thickness) eliminates this problem. In this realization, the incident wave "sees" the same geometry but with a different phase reflection, significantly decreasing the transition effects.

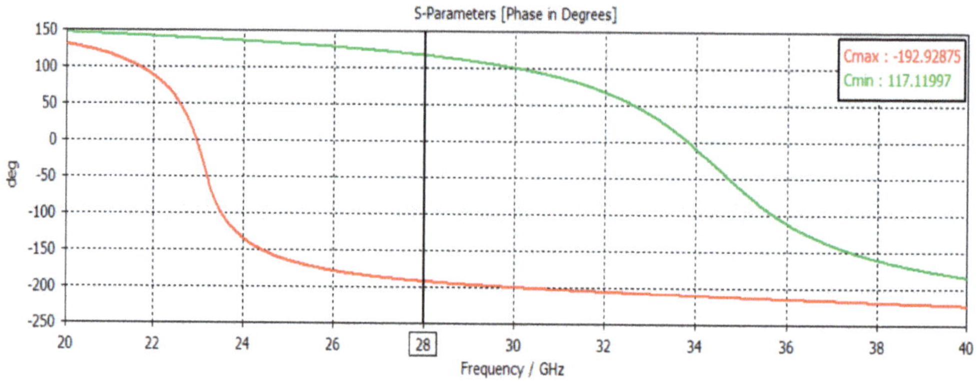

(**a**)

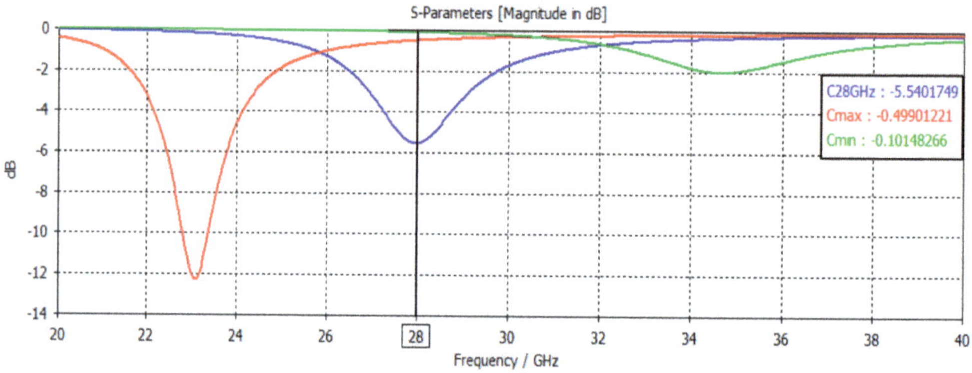

(**b**)

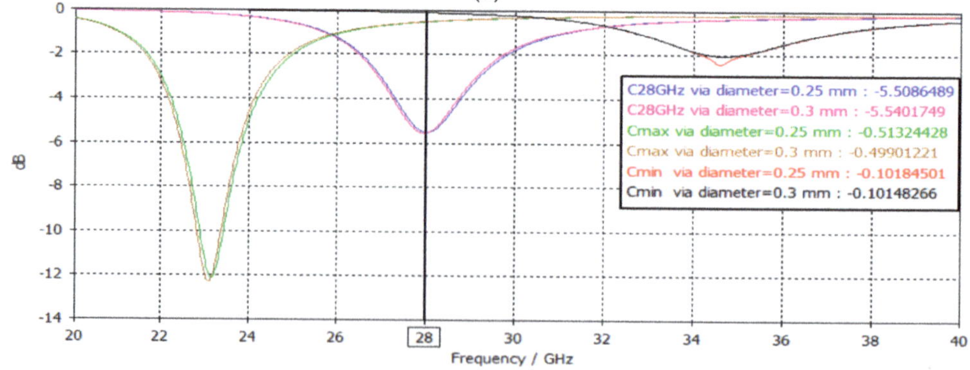

(**c**)

Figure 3. Simulation results of uniform unit cell reflection. (**a**) Phase. (**b**) Magnitude. (**c**) Influence of the via diameter.

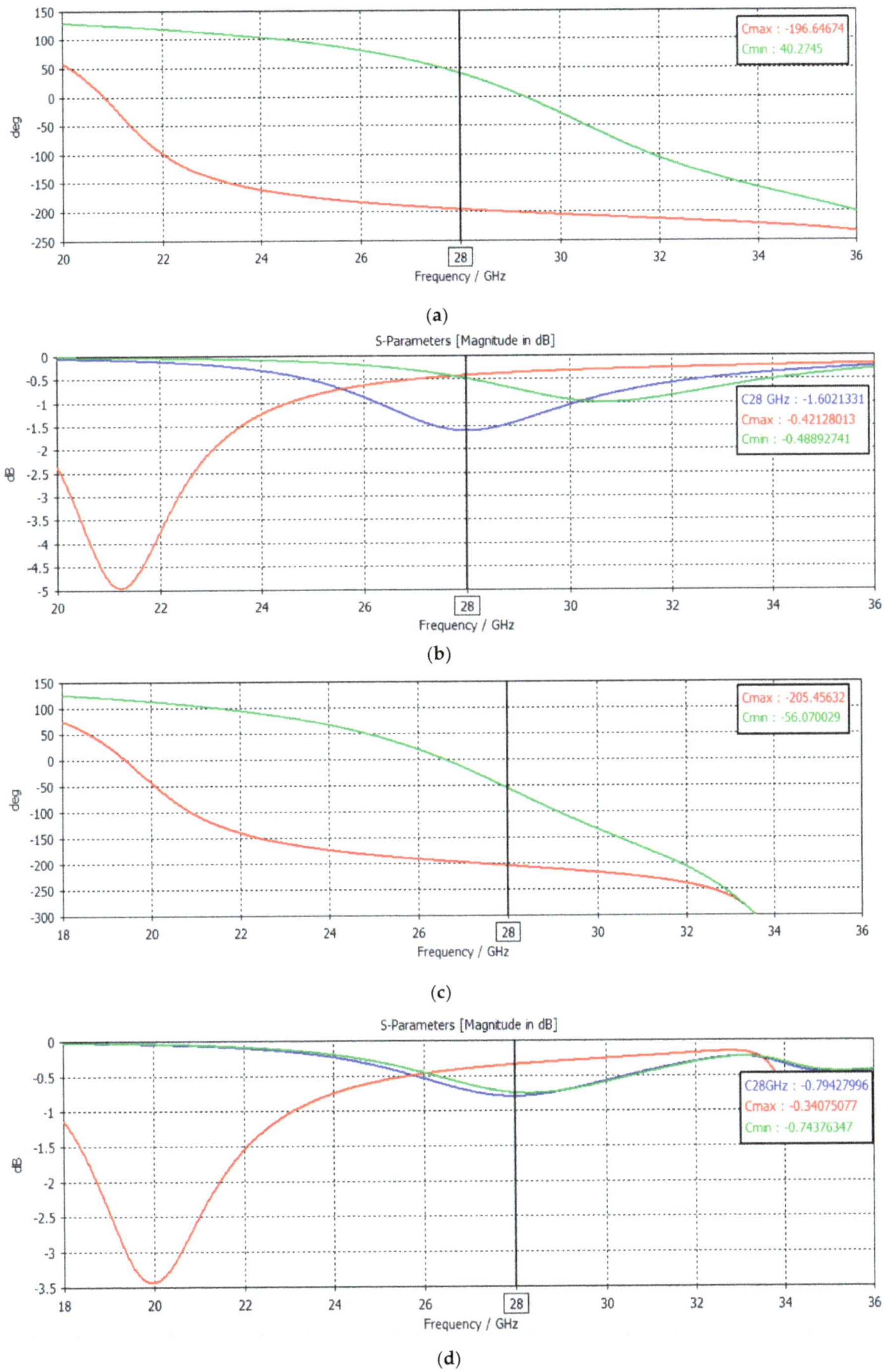

Figure 4. Unit cell reflection simulation results. (**a**) Type A—Phase. (**b**) Type A—Magnitude. (**c**) Type B—Phase. (**d**) Type B—Magnitude.

3. Metasurface Simulation Results

The MS reflector structure, including the DC bias circuit, is shown in Figure 5. The MS structure and the simplicity of DC biasing reduces the complexity and the absorption of the unit cell (Figures 3 and 4). The via and the clearance add losses to the unit cell due to the need to provide a DC bias to each varactor. The stripes of the unit cells are used as a DC ground for the varactors in one stripe column, eliminating the need for an additional via in each unit cell. The shorter strip is used for supplying the DC bias through a via, enabling each MS unit cell to be independently biased. Figure 5 describes the MS structure including the DC bias circuit where $\Delta\phi_x$ and $\Delta\phi_y$ are the reflected phase differences between two adjacent unit cells in X and Y, respectively.

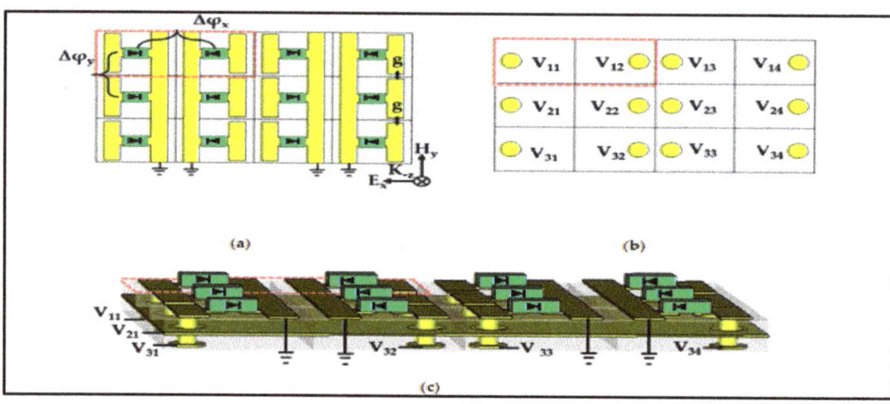

Figure 5. MS structure with varactor DC biasing method. (**a**) MS top layer. (**b**) MS back layer with varactors' DC bias indexes. (**c**) 3D inside view of the MS with the closure of the electrical circuit for each varactor [8].

The array size was 40.8 mm × 40.8 mm with 12 rows and 16 columns of unit cells, yielding 192 independent unit cells with a separate DC voltage connection, as required. Beam steering in one dimension (the X axis, in this case) requires that every unit cell in each column is connected to the same DC voltage/basis to get the desired $\Delta\phi_x$ while $\Delta\phi_y$= 0°. The steering angle θ for each angle was calculated, as shown in Table 3 using (1).

$$\theta = \sin - 1\ (\lambda\cdot\Delta\phi_x./360\cdot\Delta X) \tag{1}$$

where ΔX is the distance between the canter of two adjacent unit cells, and λ is the operating frequency wavelength. The MS was simulated using CST code for a normal incident TEM plane wave of approximately 28 GHz. The simulation results are shown in Figure 5. Results show the main lobe difference to the side lobe at more than 10 dB. The results can be improved by approximately 3 dB by using the optimal phase dynamic range with the stripes configuration. For the beam-steering mode, considering the dynamic phase range and MS size, the phase difference parameter $\Delta\phi_x$ is limited according to Equation (2) for type A and Equation (3) for type B.

$$236^\circ \geq 11 \times \Delta\phi_x \tag{2}$$

$$140^\circ \geq 7 \times \Delta\phi_x \tag{3}$$

In Figure 6a, beam steering at θ = 0° around the resonant frequency with the highest absorption is shown. The solid red line describes the beam reflected gain of one stripe (uniform) MS and the solid green line describes the beam reflected gain for the three stripe configuration. The three-stripe gain curve showed significantly better performance compared to the gain in the one-stripe configuration (red line) by approximately 3.6 dB. Furthermore, the larger the steering angle was, the better the reflected gain of the three-stripe configuration compared to one stripe (uniform) MS. Figure 7 describes the simulation results of a typical one-stripe configuration and the simulation of a three-stripe configuration for steering angles of $\theta = 5°$, $\theta = 7°$, and $\theta = 9°$.

Table 3. Parameter values in Equation (1) for beam steering (periodicity $\Delta x = 3.4$ mm).

Reflected Beam (Deg)	$\Delta\phi_x$ (Deg)
5	10
7	14
9	18

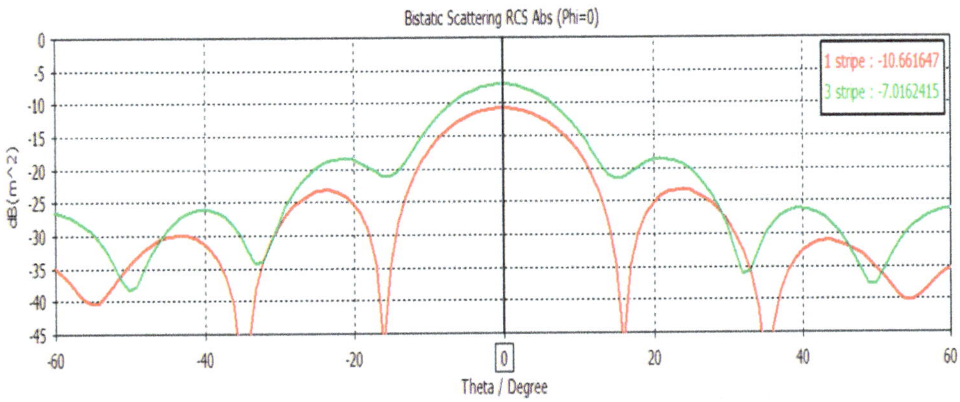

Figure 6. Beam steering far-field radar cross section (RCS) results at 28 GHz with the three-stripe configuration (green) vs. the one-stripe configuration (red). Steering angle simulations results of 0°.

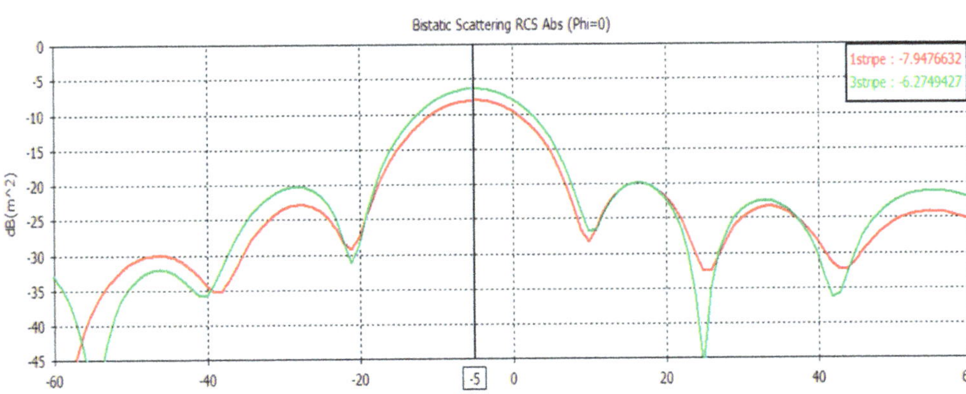

(a)

Figure 7. *Cont.*

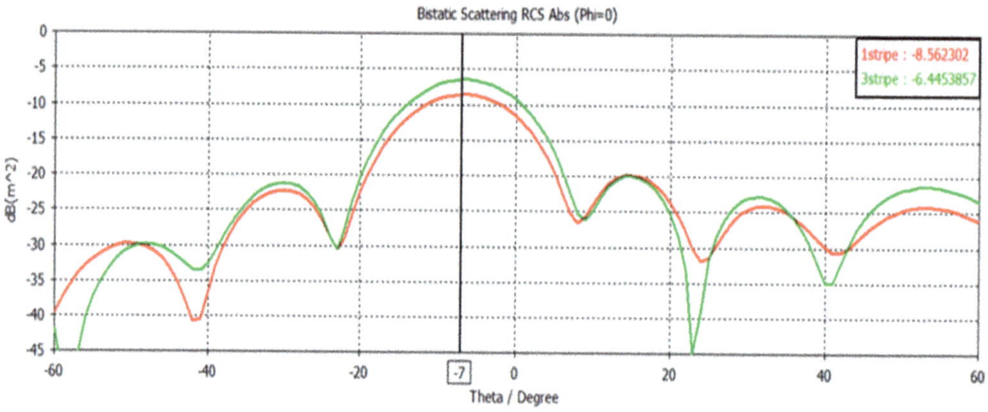

(**b**)

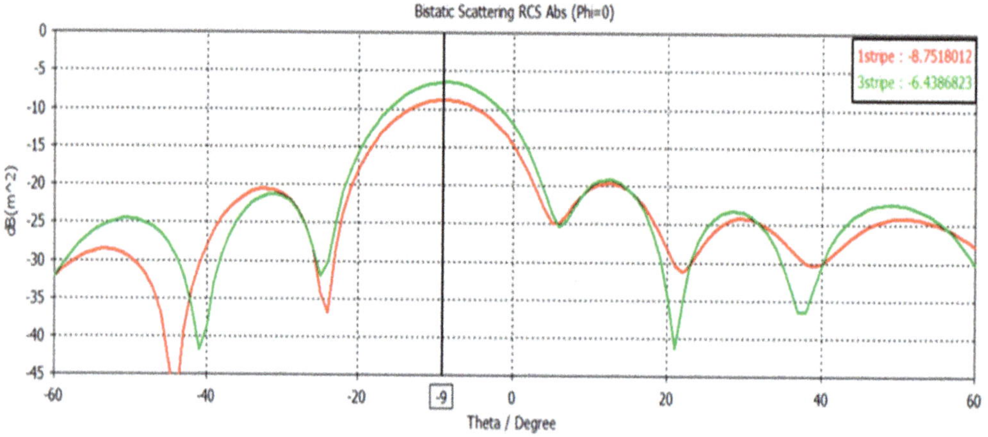

(**c**)

Figure 7. Beam steering far-field radar cross section (RCS) results at 28 GHz in the three-stripe configuration (green) vs. the one-stripe configuration (red). (**a**) Steering angle simulation results of 5°. (**b**) Steering angle simulation results of 7°. (**c**) Steering angle simulation results of 9°.

The co-polarization and cross polarization properties of the proposed three-stripe MS was investigated for the case of 9° beam steering. The simulation results show that most of the transmitted power was reflected in the co-polarization. The cross polarization was below 50 dB, referring to the co-polarization, as seen in Figure 8.

There was a correlation between the calculation in Equation (1) for beam steering and the simulation results. The stripes configuration improved the performance of the reflected beam gain by 3 dB. A significant improvement in the reflected beam angle θ for different frequencies around the operating frequency (bandwidth) was obtained with the three-stripe configuration, compared to the typical uniform configuration (one stripe). Figures 9 and 10 show the influence of a slight change in the operating frequency of the MS. The accuracy in the beam steering angle for the three-stripe configuration was about 50% better than for the typical one-stripe configuration. Far-field simulations indicated that the three-stripe configuration for a steering angle of $\theta = 10°$ showed an almost 0.7° difference (see Figure 10) compared to the uniform one-stripe configuration, which was about a 1.2–1.4° difference (see Figure 9) for the same frequencies. Table 4 summarizes the bandwidth simulation results.

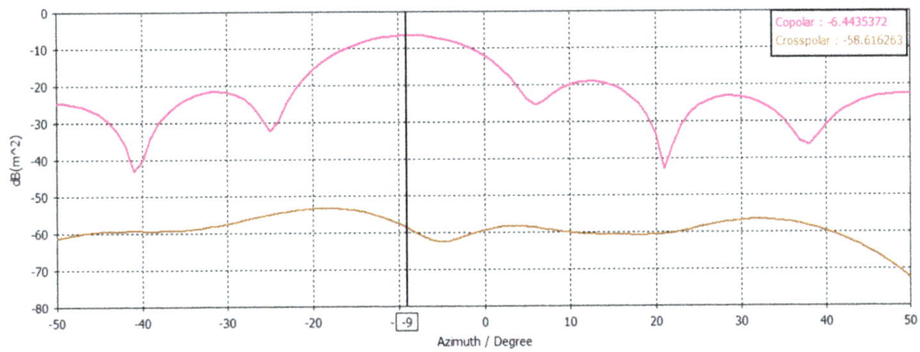

Figure 8. Results of far-field co-polarization and cross polarization for beam steering at 28 GHz in the three-stripe configuration. Steering angle simulation results of 9°.

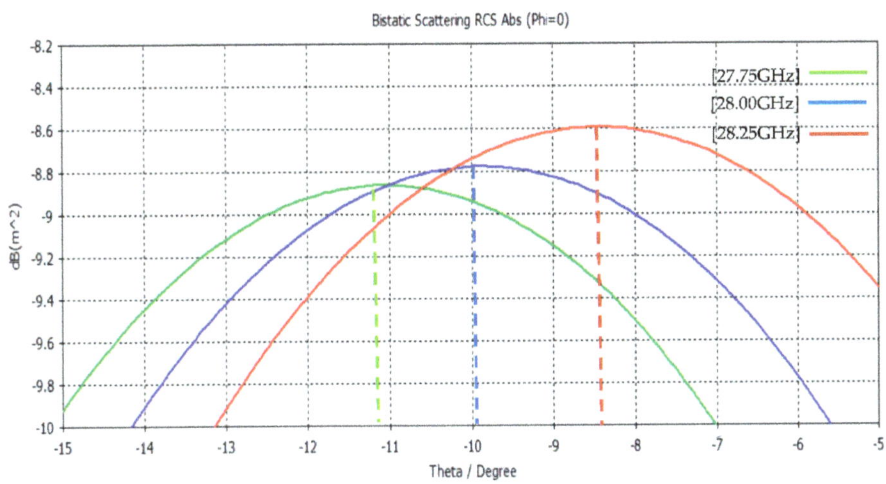

Figure 9. Results of bandwidth for beam steering 10° at 28 GHz in the one-stripe configuration.

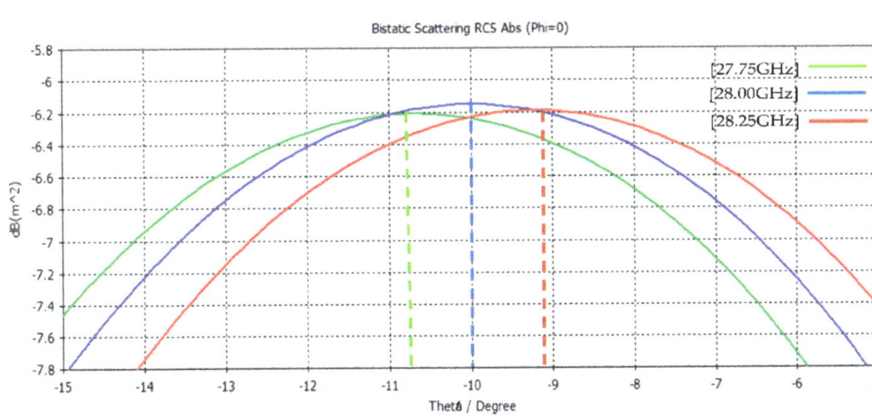

Figure 10. Results of bandwidth for beam steering 10° at 28 GHz in the three-stripe configuration.

Table 4. Results of far-field radar cross section (RCS) for beam steering around 28 GHz for a uniform configuration compared to a three-stripe configuration.

Monitor Far-Field (GHz)	Reflected Beam (Deg) One Stripe	Reflected Beam (Deg) Three Stripes
27.75	11.1	10.7
28	9.9	10
28.25	8.5	9.2

To demonstrate the advantages of the stripes method, the following combination was investigated: uniform–type A–uniform MS (three stripes method) compared to uniform MS (one stripe). Three performance parameters were simulated: reflected beam gain, SLL, and steering dynamic range. Figure 11 shows the simulation of the reflected beam for 8 deg for three stripes (solid green line) and one stripe (solid red line). The performance of the reflected beam gain and SLL of the three-stripe configuration was higher by about 3 dB.

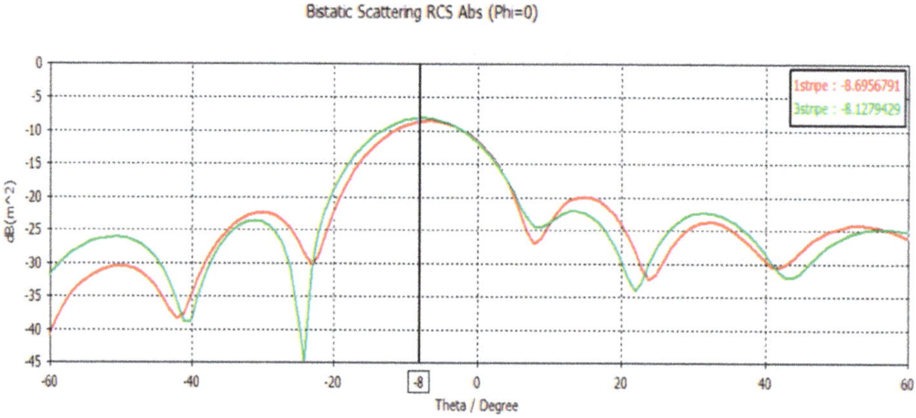

Figure 11. Beam steering far-field radar cross section (RCS) results at 28 GHz in the three-stripe (green) vs. one-stripe configuration (red). Steering angle simulation results of 8°.

The phase dynamic range for 12° and 31° are given in Figure 12. The three-stripe configuration composed of uniform–type A–uniform was able to cover a much larger dynamic range of the steering angle, as can be seen in Figure 12, whereas the type A (one stripe) configuration was unable to realize this.

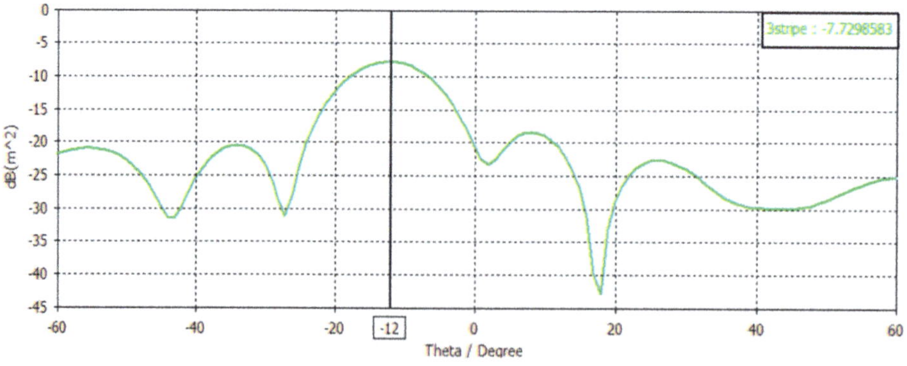

(a)

Figure 12. *Cont.*

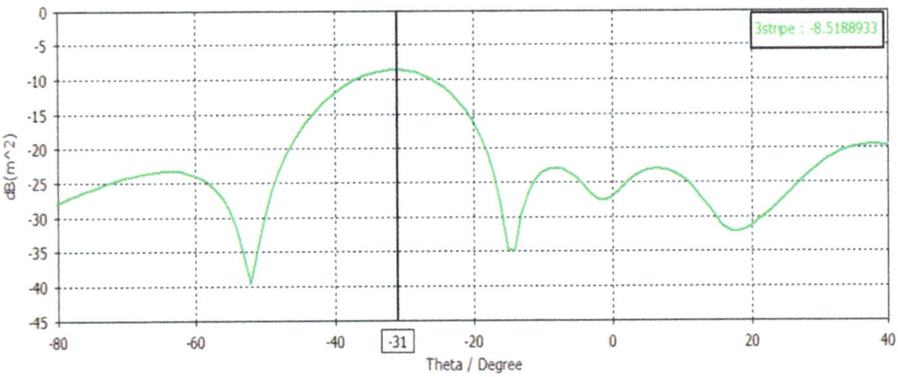

(b)

Figure 12. Beam steering far-field radar cross section (RCS) results at 28 GHz in the three-stripe configuration. (**a**) Steering angle simulation results of 12°. (**b**) Steering angle simulation results of 31°.

4. Discussion

The simulation results indicate that the three-stripe configuration proposed here yielded better performance compared to the common configuration (one stripe). Performance parameters such as reflected beam gain, SLL, beam-steering accuracy, and steering dynamic range showed better results. Figure 7 shows a 3 dB better reflected gain and SLL for the three-stripe configuration in comparison to the common one-stripe configuration. The larger the beam-steering angle, the better the reflected gain, SLL, and bandwidth in the three-stripe configuration, compared to the one-stripe configuration. The reason for this is the smaller phase dynamic range of each stripe in the three-stripe configuration (type A and type B) compared to the one-stripe configuration where the phase dynamic range was larger (Figure 4). In addition, another combination of the three-stripe configuration was investigated, in which the three-stripe configuration was again superior to the one-stripe configuration (Figures 11 and 12). The accuracy of the reflected angle for different frequencies around the operating frequency (bandwidth) was found to be better for the three-stripe configuration compared to the one-stripe configuration, as shown in Table 4. A 50% better accuracy was demonstrated (Figures 9 and 10).

The construction of a three-stripe MS could be realized by using a plastic frame holder in the front and a plastic board on the back side for mechanical strength, as shown in Figure 13. In Figure 13, the incident beam "sees" the whole MS uniformly, aligned at Z = 0 (the front side of the MS). On the back side, it is different, but no effect on the reflected beam was found.

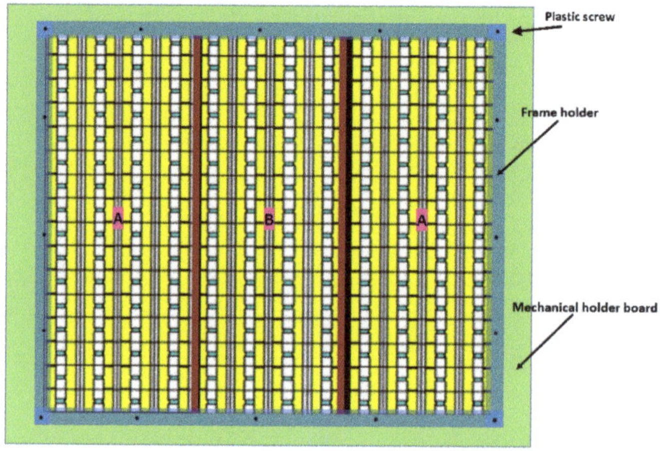

Figure 13. Full and detailed realization of the three-stripe MS.

5. Conclusions

In this paper we demonstrated a new concept of a metasurface reflector for 5G communication based on a stripe configuration. This unit cell offers beam steering in the Ka-band with low losses and better beam-steering performance. The unit cell design fits the standard PCB technology. In general, the phase dynamic range of the unit cell and its losses were inversely related. The use of the three-stripe configuration led to losses lower than 1.5 dB at 28 GHz for Type A (Figure 2) and losses lower than 0.8 dB at 28 GHz for Type B (see Figure 2), compared to uniform unit cell losses of about 5.5 dB at the same frequency, as shown in Figures 3 and 4. The three-stripe configuration enabled superior beam steering with better reflected beam gain in comparison to to the one-stripe configuration. Furthermore, the improvement of SLL and the reflected gain was about 3 dB, and the improvement of the reflected beam angle accuracy was about 50%.

Author Contributions: Conceptualization, E.R. and D.R.; methodology, E.R., D.R., and A.A.; software, E.R.; validation, E.R.; data curation, E.R.; formal analysis, E.R. and D.R.; investigation, E.R. and D.R.; resources, E.R., D.R., and A.A.; writing, E.R. and D.R.; writing—review and editing, E.R., D.R., and A.A.; visualization, E.R. and D.R.; supervision, A.A. and D.R.; project administration, E.R.; funding acquisition, A.A. All authors have read and agreed to the published version of the manuscript.

Funding: This research received no external funding.

Conflicts of Interest: The authors declare no conflict of interest.

References

1. Veselago, V.; Narimanov, E. The left hand of brightness: Past, present and future of negative index materials. *Nat. Mater.* **2006**, *5*, 759–762. [CrossRef] [PubMed]
2. Nemat-Nasser, S.C.; Amirkhizi, A.V.; Padilla, W.J.; Basov, D.N.; Nemat-Nasser, S.; Bruzewicz, D.; Whitesides, G. Terahertz plasmonic composites. *Phys. Rev. E* **2007**, *75*, 036614. [CrossRef] [PubMed]
3. Syms, R.R.A.; Shamonina, E.; Kalinin, V.; Solymar, L. A theory of metamaterials based on periodically loaded transmission lines: Interaction between magnetoinductive and electromagnetic waves. *J. Appl. Phys.* **2005**, *97*, 064909. [CrossRef]
4. Landy, N.I.; Sajuyigbe, S.; Mock, J.J.; Smith, D.R.; Padilla, W.J. Perfect Metamaterial Absorber. *Phys. Rev. Lett.* **2008**, *100*, 207402. [CrossRef] [PubMed]
5. Schurig, D.; Mock, J.J.; Justice, B.J.; Cummer, S.A.; Pendry, J.B.; Starr, A.F.; Smith, D.R. Metamaterial Electromagnetic Cloak at Microwave Frequencies. *Science* **2006**, *314*, 977–980. [CrossRef] [PubMed]
6. Gil Litmanovitch, G.L.; Rrotshild, D.R.D.; Abramovich, A. Flat mirror for millimeter-wave and terahertz imaging systems using an inexpensive metasurface. *Chin. Opt. Lett.* **2017**, *15*, 011101–11105. [CrossRef]

7. Rotshild, D.; Abramovich, A. Realization and validation of continuous tunable metasurface for high resolution beam steering reflector at K-band frequency. *Int. J. RF Microw. Comput. Eng.* **2021**, *31*. [CrossRef]
8. Rotshild, D.; Rahamim, E.; Abramovich, A. Innovative Reconfigurable Metasurface 2-D Beam-Steerable Reflector for 5G Wireless Communication. *Electronics* **2020**, *9*, 1191. [CrossRef]
9. Veselago, V.G. Energy, linear momentum and mass transfer by an electromagnetic wave in a negative-refraction medium. *Phys. -Uspekhi* **2009**, *52*, 649. [CrossRef]
10. Pendry, J.B.; Holden, A.J.; Robbins, D.J.; Stewart, W.J. Magnetism from Conductors and Enhanced Nonlinear Phenomena. *IEEE Trans. Microw. Theory Tech.* **1999**, *47*, 2075. [CrossRef]
11. Smith, D.R.; Padilla, W.; Vier, D.C.; Nemat-Nasser, S.C.; Schultz, S. Composite Medium with Simultaneously Negative Permeability and Permittivity. *Phys. Rev. Lett.* **2000**, *84*, 4184–4187. [CrossRef] [PubMed]
12. Aobo, L.; Shreya, S.; Sievenpiper, D. Metasurfaces and their applications. *Nanophotonics* **2018**, *7*, 989–1011.
13. Rotshild, D.; Azoulay, Y.; Ochana, M.; Shulzinger, A.; Abramovich, A. Real time detection and recognition of micro-poisons in aqueous solutions and atmosphere using perfect absorber metamaterial in millimeter wavelength regime. *Head Neck.* **2015**, 15678708. [CrossRef]
14. Solderable AlGaAs Flip Chip PIN. Available online: https://cdn.macom.com/datasheets/MADP-000907-14020x.pdf (accessed on 11 June 2020).
15. Tian, W.; Li, P.; Yuan, L. Research and Analysis of MEMS Switches in Different Frequency Bands. *Micromachines* **2018**, *9*, 4185. [CrossRef] [PubMed]
16. Foo, S. Liquid-crystal-tunable metasurface antennas. In Proceedings of the 2017 11th European Conference on Antennas and Propagation (EUCAP), Paris, France, 19–24 March 2017.
17. Mavridou, M.; Feresidis, A.P. Dynamically reconfigurable high impedance and frequency selective metasurfaces using piezoelectric actuators. *IEEE Trans. Antennas Propag.* **2016**, *64*, 5190–5197. [CrossRef]
18. Yaghmaee, P.; Karabey, O.H.; Bates, B.; Fumeaux, C.; Jakoby, R. Electrically Tuned Microwave Devices Using Liquid Crystal Technology. *Int. J. Antennas Propag.* **2013**, *2013*, 824214. [CrossRef]
19. MACOM AVR-011020-1411, Solderable GaAs Constant Gamma Flip-chip Varactor Diode. Available online: https://cdn.macom.com/datasheets/MAVR-011020-1141.pdf (accessed on 11 June 2020).
20. Rotshild, D.; Abramovich, A. Ultra-Wideband reconfigurable X-band and Ku-band metasurface beam-steerable reflector for satellite communications. *Electronics* **2021**, *10*, 2165. [CrossRef]
21. Rogers Corporation. RT/Duroid®5880. Available online: https://www.rogerscorp.com//media/project/rogerscorp/documents/advanced-connectivity-solutions/english/data-sheets/rt-duroid-5880lz-high-frequency-laminates.pdf (accessed on 11 June 2020).
22. Wang, M.; Ma, H.F.; Tang, W.X.; Zhang, H.C.; Wang, Z.X.; Cui, T.J. Programmable Controls of Multiple Modes of Spoof Surface Plasmon Polaritons to Reach Reconfigurable Plasmonic Devices. *Adv. Mater. Technol.* **2019**, *4*, 1800603. [CrossRef]
23. Wang, M.; Ma, H.F.; Tang, W.X.; Zhang, H.C.; Jiang, W.X.; Cui, T.J. A Dual-Band Electronic-Scanning Leaky-Wave Antenna Based on a Corrugated Microstrip Line. *IEEE Trans. Antennas Propag.* **2019**, *67*, 3433–3438. [CrossRef]
24. Rütschlin, M.; Wittig, T.; Iluz, Z. Phased antenna array design with CST Studio Suite. In Proceedings of the 2016 10th European Conference on Antennas Propag (EuCAP), Davos, Switzerland, 10–15 April 2016; IEEE: Davos, Switzerland, 2016; pp. 1–5.
25. Loh, T.H. High Impedance Surface Electromagnetic Band Gap Metamaterials: Design Approach and Applications for Antenna Engineering ARMMS, Northampton, UK, November 2011. Available online: https://www.armms.org/media/uploads/15_armms_nov11_thongloh.pdf (accessed on 15 May 2021).

applied sciences

MDPI

Article

Quarter Wavelength Fabry–Perot Cavity Antenna with Wideband Low Monostatic Radar Cross Section and Off-Broadside Peak Radiation

Hassan Umair [1], Tarik Bin Abdul Latef [1,*], Yoshihide Yamada [2], Tayyab Hassan [3], Wan Nor Liza Binti Wan Mahadi [1], Mohamadariff Othman [1], Kamilia Kamardin [2] and Mousa I. Hussein [4]

1 Department of Electrical Engineering, University of Malaya, Kuala Lumpur 50603, Malaysia; hsn.umr@gmail.com (H.U.); wnliza@um.edu.my (W.N.L.B.W.M.); mohamadariff@um.edu.my (M.O.)
2 Malaysia-Japan International Institute of Technology (MJIIT), Universiti Teknologi Malaysia, Kuala Lumpur 54100, Malaysia; yoshihide@utm.my (Y.Y.); kamilia@utm.my (K.K.)
3 Centre of Excellence in Science and Applied Technologies (CESAT), Islamabad 44000, Pakistan; tayyabamer@hotmail.com
4 Department of Electrical Engineering, United Arab Emirates University, Al Ain 15551, UAE; mihussein@uaeu.ac.ae
* Correspondence: tariqlatef@um.edu.my

Abstract: Since antennas are strong radar targets, their radar cross section (RCS) reduction and radiation enhancement is of utmost necessity, particularly for stealth platforms. This work proposes the design of a Fabry–Perot Cavity (FPC) antenna which has wideband low monostatic RCS. While in the transmission mode, not only is gain enhancement achieved, but radiation beam is also deflected in the elevation plane. Moreover, the design is low-profile, i.e., the cavity height is ~$\lambda/4$. A patch antenna designed at 6 GHz serves as the excitation source of the cavity constructed between the metallic ground plane and superstrate. The superstrate structure is formed with absorptive frequency selective surface (AFSS) in conjunction with dual-sided partially reflective surface (PRS). Resistor loaded metallic rings serve as the AFSS, while PRS is constructed from inductive gradated mesh structure on one side to realize phase gradient for beam deflection; the other side has fixed capacitive elements. Results show that wideband RCS reduction was achieved from 4–16 GHz, with average RCS reduction of about 8.5 dB over the reference patch antenna. Off-broadside peak radiation at −38° was achieved, with gain approaching ~9.4 dB. Simulation and measurement results are presented.

Keywords: Fabry–Perot Cavity (FPC); radar cross section; frequency selective surface; partially reflecting surface; off-broadside peak radiation

Citation: Umair, H.; Latef, T.B.A.; Yamada, Y.; Hassan, T.; Mahadi, W.N.L.B.W.; Othman, M.; Kamardin, K.; Hussein, M.I. Quarter Wavelength Fabry–Perot Cavity Antenna with Wideband Low Monostatic Radar Cross Section and Off-Broadside Peak Radiation. *Appl. Sci.* **2021**, *11*, 1053. https://doi.org/10.3390/app11031053

Received: 1 December 2020
Accepted: 7 January 2021
Published: 25 January 2021

1. Introduction

Stealth platforms have low radar cross section (RCS), but their radar signature increases significantly when antennas are mounted on them for communication purposes [1,2]. This can compromise their ability to counter the radar waves, so, in this regard, design and development of low RCS antennas is deemed necessary, for safety and security.

Reduction of the antenna's RCS is a critical feat, and several methods have been investigated to ensure that the antenna radiation properties are least affected while attempting to reduce its RCS. One of the methods is structural/geometrical shaping [3–7], in which the shape of the radiating structure is modified to ensure the backscatter avoids the threat direction. The other method is based on periodic structures, and this includes the use of radar absorbing materials (RAMs) [8–10], frequency selective surface (FSS) ground plane [11,12], FSS radome [13,14], electromagnetic bandgap (EBG) structures [15–17], artificial magnetic conductors (AMCs) [18], perfect metamaterial absorbers (MAs) [19–22], and polarization conversion metasurfaces (PCMs) [23,24]. In all of the above configurations, the periodic structure is implemented either at the ground plane, on top of the radiator, or loaded around the planar radiator, and, in all of these implementations, although the RCS is lowered, the antenna radiation property either remains just intact, or it slightly deteriorates.

To improve the antenna radiation properties in parallel with lowering the RCS, further research has led to using partially reflecting surfaces (PRSs) in a Fabry–Perot Cavity (FPC) configuration, as evident in [25–28]. In all of these works, backscatter reduction was achieved, and broadside antenna gain was enhanced. To achieve this, mainly an absorptive FSS (AFSS) (periodic loop elements with lumped resistors) surface was used with a uniform PRS—a PRS employing identical unit cell elements in a grid, and hence identical transmission/reflection responses over the entire surface of the superstrate. With reference to antenna, in transmission mode, the uniform superstrate acts as a PRS for broadside gain enhancement, and in receiving mode, it acts as an electromagnetic (EM) absorber for normal EM wave illumination over a wide frequency range.

Some interesting work can evolve if a PRS that has a phase gradient can be used with an AFSS after appropriate design adjustments. Our aim in this work is to develop a superstrate structure that consists of phase gradient metasurface (PGM) conjoined with an AFSS, such that not only the wideband monostatic RCS reduction and peak gain enhancement can be achieved for a patch antenna, but also the peak radiation can be steered in a fixed angle, which becomes an additional antenna functionality in comparison to the works done previously. In addition, the design goal also includes the realization of a reduced cavity profile. In [29], the use of PGM can be found, integrated with an AFSS. Its primary purpose there is to scatter away the in-band incident wave; however, the presence of an absorptive surface may only serve the same to an appropriate level, by suppressing the in-band incident wave. The peak radiation direction is still towards broadside.

This study investigates the use of a PRS constructed with unit cells that have phase shifts implemented by a dimensional gradient, and henceforth progressively varying transmission/reflection properties over the surface of the superstrate. The designed PRS is a composite structure, meaning thereby it utilizes both sides of the dielectric, and this feature aids in reducing the cavity height to $\lambda/4$. The absorptive surface consists of periodic loop elements loaded with lumped resistors, and it works in conjunction with the composite PRS that is formed by gradated mesh (inductive) structure on one side and constant patch (capacitive) elements on the other side. The excitation source of the cavity is a patch radiator designed at 6 GHz (C-band). Simulations have been validated with a fabricated prototype. Off-broadside peak gain of 9.4 dB was achieved at -38° offset in the elevation plane, along the axis of the gradient. The wideband (in-band + out-of-band) monostatic RCS reduction was achieved over a frequency range of 4–16 GHz (120%), with average RCS reduction exceeding 8.5 dB for the two orthogonal (x/y) polarizations.

The low scattering property of the proposed antenna makes it suitable to be integrated with stealth type platforms. That is because the platform's low observability would still remain low despite mounting the antenna onto it for communication, and this paves the way for its multiple applications in the military and defense realm. One example is a side looking air borne radar (SLAR) [30], where the antenna points to a sideward direction and requires physical tilting of its structure. The proposed antenna can potentially be used in this scenario. Similarly, for the aerial security, surveillance, and reconnaissance applications, it can be used on unmanned aerial vehicles (UAVs) and drones where downward pointing high gain beam is more pragmatic for communication with the ground targets [31]. The development of low RCS multiple-input multiple-output (MIMO) antennas is also becoming popular due to their technological advantages, and as such, the proposed technique can be further developed to realize pattern decorrelated low observable MIMO antennas [32,33]. In addition, the antenna can be utilized for any military communication application where fix tilt-angled communication is required [34].

2. Unit Cell Design and Proposed FPC Antenna

The goal is to design a stacked combination of layers of FSS elements, in a unit cell configuration, which when paced as a grid above the patch antenna (excitation source) in an FPC configuration, should realize the following four functionalities/objectives in parallel:

1. Wideband monostatic RCS reduction.
2. Appropriate gain enhancement.
3. Off-broadside peak radiation.
4. Reduced cavity height ($\lambda/4$).

Where λ is free space wavelength at operating frequency. Conventionally, the FP cavity, with an excitation source within, resonates when a PRS is placed at a height of ~$\lambda/2$ above the ground plane reflector, and results in an enhanced gain radiation of the source antenna. The cavity height (h) at wavelength (λ) corresponding to the operating frequency can be calculated as [35]:

$$h = (\varphi_{PRS} + \varphi_G)\,\lambda/4\pi + N\lambda/2, \quad (1)$$

where φ_{PRS} is the phase of reflection coefficient of PRS, φ_G is reflection phase of ground reflector, and N defines the resonance order. φ_G is further estimated as:

$$\varphi_G = \pi - 2\tan^{-1}(Z_d \tan(\beta d)/Z_0), \quad (2)$$

where Z_d and Z_0 are characteristic impedances of the dielectric and air, respectively, β represents the dielectric phase constant (given as $2\pi/\lambda$), and d is the dielectric thickness over which the ground plane is lying. For the metallic reflector only (i.e., without the dielectric), the reflection phase is π. For $N = 0$, the resonant cavity height turns out to be $\lambda/2$ if φ_{PRS} is assumed to be of π rad. If a PRS can be designed to exhibit a 0 reflection phase, the cavity height can be reduced to $\lambda/4$.

The unit cell with port designations is shown in Figure 1a. To obtain wideband RCS reduction, the top surface of the unit cell, as shown in Figure 1b, was constructed from a closed metallic ring of square shape, with four RF resistors (100 Ω each) soldered on four sides. Such a resistive periodic surface (AFSS) mounted on a perfect electric conductor (PEC) plane with a dielectric sandwiched in-between serves as a wideband RAM for an impinging electromagnetic (EM) wave [36]. However, for the proposed objectives, the backing PEC plane has to be replaced with an appropriate PRS, so that all our four objectives can be simultaneously achieved. A survey of literature [37–39] shows various designs of reflective surfaces; however, the appropriate PRS to be conjoined with the selected AFSS should be:

1. Symmetric in design, so that wideband RCS reduction can be achieved for both polarizations of the incident radar wave, i.e., transverse electric (TE) and transverse magnetic (TM), and
2. It should give 0 reflection phase so that once mounted above patch antenna, the cavity height can be reduced to $\lambda/4$.

Such a surface can be designed if a dual-sided PRS dielectric is constructed with an inductive mesh (aperture) grid on its top and a capacitive patch grid on its bottom [40]. AFSS of periodic square rings when backed by such a surface would achieve wideband RCS reduction, gain enhancement, and reduce cavity height; however, to also achieve radiation beam deflection, a gradient of phase has to be implemented within this composite PRS. To achieve this, a gradation in the size of mesh aperture was kept, as an inductive gradient achieves higher beam deflection than a capacitive gradient [40]. The proposed constituent unit cell elements of the PRS are shown in Figure 1c,d. Therefore, the designed final unit cell consists of resistive ring on the top, followed by air gap, and followed by gradated aperture on top of a constant patch element. However, intuitively, below the resistive ring the presence of a constant patch on top of a gradated mesh would have been more suitable, in that the variation of mesh aperture would have least affected the desired absorption frequency response. In fact, initial unit cell simulations were performed with that configuration; however, two problems arose:

1. The achieved reflection phase gradient was meagre and seemed insufficient to achieve significant beam tilt.
2. The reflection phase values were not supportive of reduced cavity height.

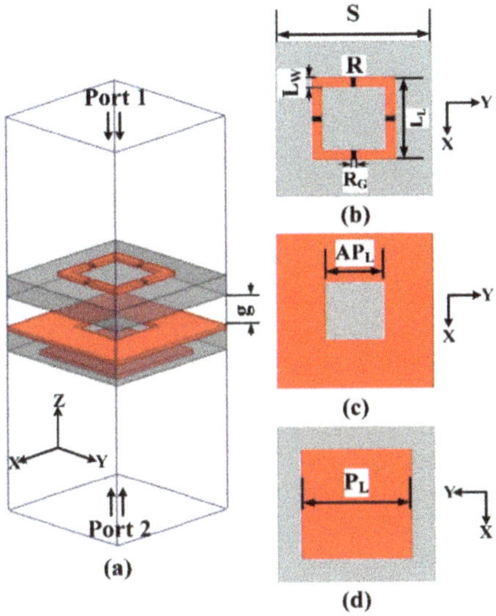

Figure 1. Unit cell structure. (**a**) Perspective view. (**b**) Top side. (**c**) Middle layer. (**d**) Bottom side. (g = 3.2 mm, S = 15 mm, L_W = 0.9 mm, L_L = 8 mm, R_G = 0.5 mm, R = 100 Ω, AP_L = variable, P_L = 10.5 mm).

Henceforth, the configuration of the PRS was flipped below the AFSS (gradient mesh above constant patch), and interestingly, the unit cell simulations showed encouraging results for fulfilling all objectives, and are discussed next in detail.

The high frequency structure simulator (HFSS) unit cell parametric simulations employing periodic boundaries and Floquet ports were performed to compute the finalized scattering (S) parameters. While optimizing the S-parameters, the following guidelines were followed [25,28]:

- For incoming wave absorption (port 1 to port 2), reflection (S_{11}) magnitude as well as transmission (S_{21}) magnitude had to be below −10 dB over a wide range of frequencies, to achieve at least 80% of incident wave absorption.
- In the transmission mode (port 2 to port 1), reflection coefficient (S_{22}) had to show high partial reflectivity as well as progressive phase over the gradated apertures, at operating frequency, to achieve high gain as well as off-broadside radiation.

The plots in Figure 2 are of TE (transverse electric) wave polarization, and it is important to note that due to the symmetry of the unit cell structure, the plots for TM (transverse magnetic) wave polarization are expected to be similar as well, and hence have not been shown here. The co-/cross-polarized reflection and transmission plots for a wave incident towards −z-axis are shown in Figure 2a,b, respectively. Solid lines depict the co-polarized components, while the dashed lines represent the cross-coupled components. Different curves correspond to the varying aperture sizes (AP_L: 2 mm to 10 mm). For the co-polarized S_{11} and S_{21} responses shown in Figure 2a,b, it can be seen that starting from 7 GHz (out-of-band) and extending well into high frequency region, more than 80% absorption (A) ($A = 1 - |S_{11}|^2 - |S_{21}|^2$) is being achieved for all aperture sizes. This owes to the values of the incoming wave reflection and transmission, which are below

−10 dB over a wideband. S_{11} and S_{21} magnitudes around operating frequency (6 GHz) show partial absorption, which means a reasonable extent of RCS reduction should occur at in-band frequencies also. For the cross-coupled S_{11} and S_{21} responses shown in Figure 2a,b, it is evident that their magnitudes are significantly low, hence establishing the efficiency and purity of absorption.

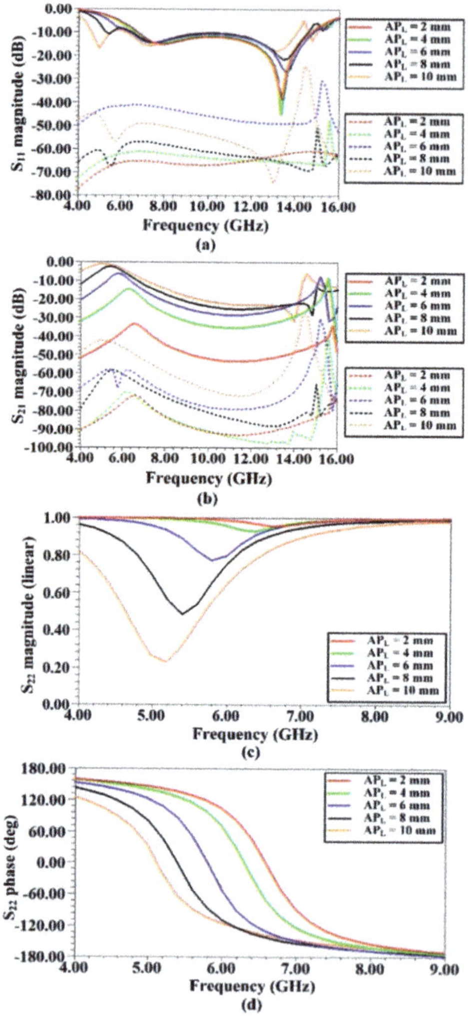

Figure 2. Unit cell scattering parameters for ±z-axis incident wave (TE polarization). (**a**) S_{11} magnitude response. (**b**) S_{21} magnitude response. (**c**) S_{22} magnitude (linear) response. (**d**) S_{22} phase response. (Note: The dashed traces shown in (**a**,**b**) represent cross-coupled/polarized reflection and transmission components).

S_{22} magnitude response is shown in Figure 2c for gradating aperture values. At operating frequency, the reflection magnitudes lie between 0.64–0.99 (linear). Figure 2d depicts the phase response of S_{22}. It shows that the reflection phase values range between 104° to −114° over aperture variation of 2 mm to 10 mm, realizing a significant phase gradient at 6 GHz. Furthermore, the extents of phase range closely follow ±90°, a reflection phase range criteria of the PRS to reduce the cavity height to $\lambda/4$ [40]. From Figure 2c,d, it can be inferred that an off-broadside radiation with a high gain can be realized in the FPC configuration.

The reflection/transmission characteristics of the unit cell for oblique angle incidences are shown in Figure 3, and the results are given for both TE and TM wave polarizations. All results were plotted for AP_L = 5.5 mm (the center of the gradient, see Table 1). From Figure 3a,b depicting TE wave polarization, it is evident that despite the variation of the angle of incidence, low magnitudes (less than −10 dB) of reflection as well as transmission are still being obtained over a wide band (starting from about 7 GHz and beyond). For the TM case, a similar behavior is also observed, and can be witnessed from Figure 3c,d. It can be asserted that the PRS backed AFSS, as an absorber, shows good angular stability.

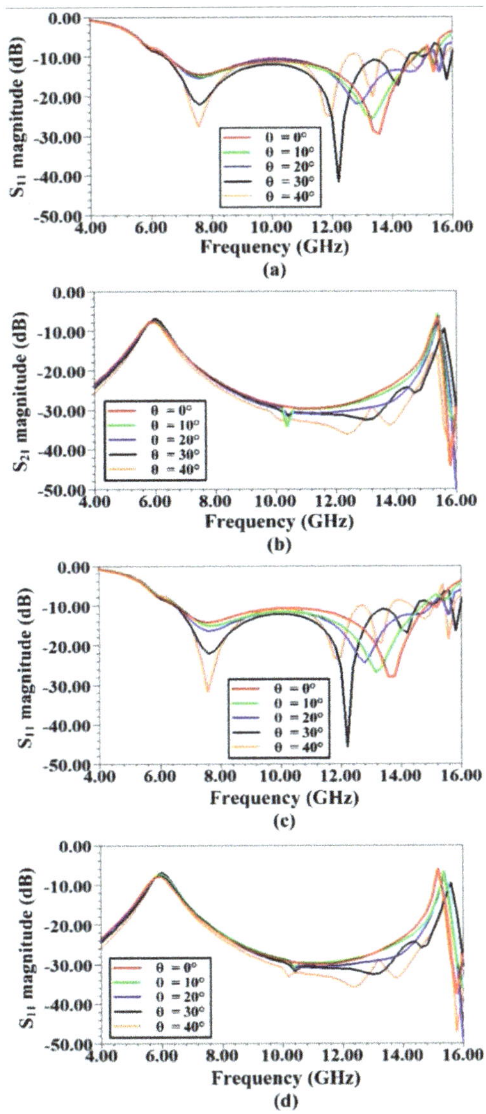

Figure 3. Unit cell response to oblique angle incidences of TE/TM wave polarizations. (**a**) Reflection response for TE wave. (**b**) Transmission response for TE wave. (**c**) Reflection response for TM wave. (**d**) Transmission response for TM wave. (All results shown for AP_L = 5 mm).

Table 1. Gradient aperture lengths (AP_L) and corresponding reflection phases.

AP_L (mm)	2.4	3.5	5.1	5.5	5.8	6.2	6.6
Reflection Phases	97.72°	86.24°	10.56°	−23.89°	−38.38°	−63.91°	−81.57°

A 7 × 7 unit cell array was mounted on top of a rectangular patch antenna, as shown by the simulated model in Figure 4a, and experimental prototype in Figure 4b. Cavity height (h) is 13 mm (~λ/4). The coaxial feed offset is 3.2 mm from the patch center towards +x-axis. All dielectric laminates are of Rogers RO4003C material (ε_r = 3.55 (design), thickness 1.52 mm). Gradient implementation is along y-axis, with the AP_L values and the corresponding reflection phases shown in Table 1.

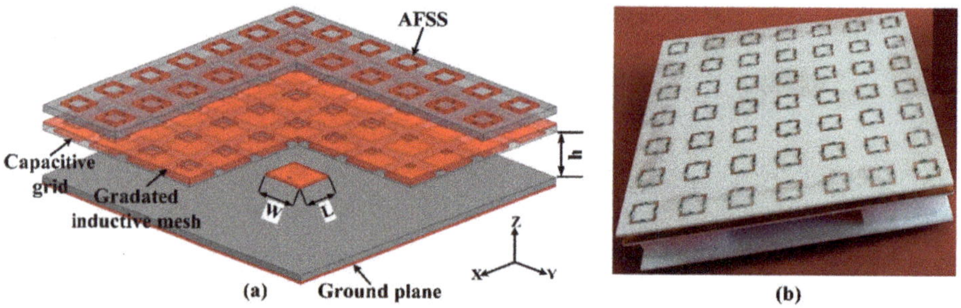

Figure 4. Proposed antenna. (**a**) Simulated model (cutout view). (**b**) Fabricated prototype. (h = 13 mm, L = 12 mm, W = 14 mm).

3. Simulation and Experimental Results

Simulated vs. measured S_{11} response of the proposed antenna is shown in Figure 5a. Sharp resonance was achieved at 6 GHz, and impedance bandwidth (−10 dB) is 443 MHz (7.26%). Gain vs. frequency plot of the proposed antenna is also shown in Figure 5a. At operating frequency, peak gain achieved is ~9.4 dB. Gain bandwidth (3 dB) is from 5.82 GHz to 6.08 GHz (4.37%). A satisfactory agreement exists between simulated and measured results, establishing the accuracy of simulation and fabrication. Figure 5b illustrates the S_{11} and gain vs. frequency response of the reference antenna (antenna without the superstrate assembly). It can be seen that −10 dB resonance of the reference antenna is slightly right-shifted as compared to the resonance of the proposed antenna. This frequency shifting is attributed to the input impedance variation when the dielectric superstrate is absent. Meanwhile, the gain of the proposed antenna is 6.4 dB at operating frequency, and is increased by 3 dB in the presence of the superstrate assembly. Figure 6a illustrates the far field H-plane radiation pattern at 6 GHz. Off-broadside radiation at an angle of −38° was achieved, deflected in the elevation plane, with side lobe levels (SLLs) ~10 dB below the maximum. Deflection angle is aligned to the axis in which the increase of aperture size (gradient) is implemented. E-plane pattern is shown in Figure 6b. Manufacturing tolerances as well as positioning errors during measurements can be the cause of difference between simulated and measured patterns. The far field H-plane and E-plane radiation patterns of the reference antenna are shown in Figure 6c,d, respectively. It can be witnessed that without the superstrate, the antenna radiates towards broadside. The radiation pattern plotted at various frequencies is shown in Figure 7. The pattern is satisfactorily uniform over a bandwidth of 50 MHz, although a little deterioration of the SLLs occurs.

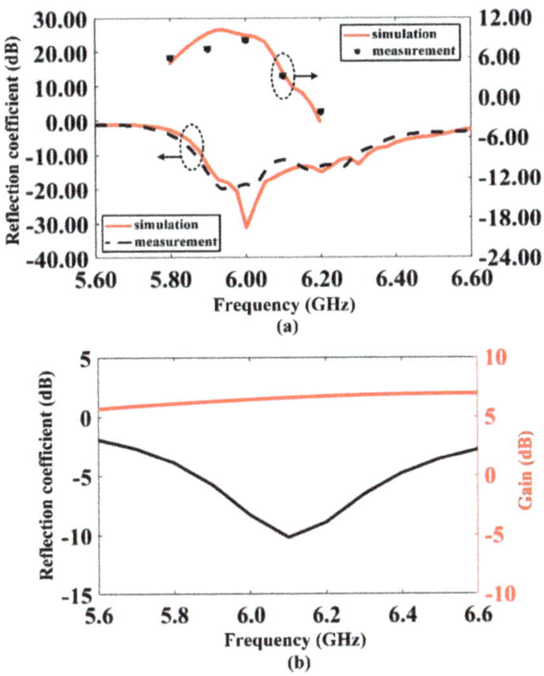

Figure 5. Reflection coefficient and gain frequency response. (**a**) Proposed antenna. (**b**) Reference antenna.

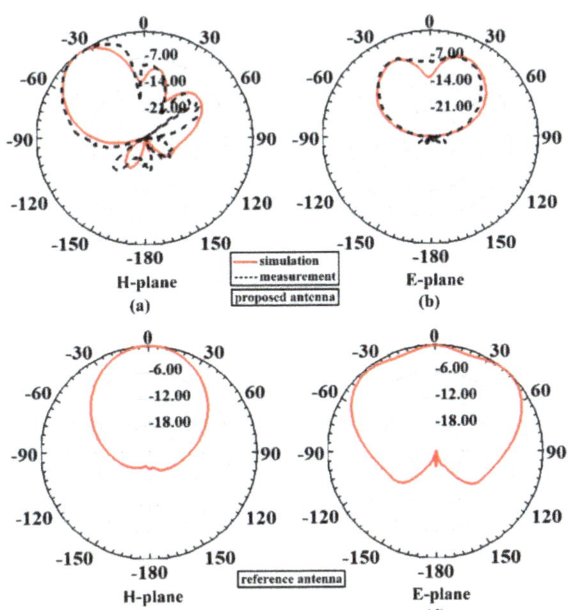

Figure 6. Radiation patterns of the proposed antenna, (**a**) H-plane. (**b**) E-plane. Radiation patterns of the reference antenna, (**c**) H-plane. (**d**) E-plane.

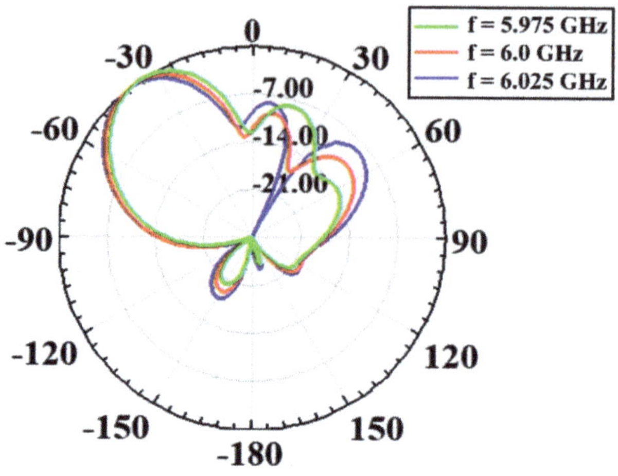

Figure 7. Radiation beam plots (H-plane) at three different frequencies (f).

The reference antenna's radiation efficiency at operating frequency is 95%; however, the proposed antenna's efficiency diminishes to 60%. The primary reason for this efficiency reduction can be deduced from Figure 2b, where the S_{21} transmission is shown. It notes that since the unit cell/superstrate is a passive structure, its S_{12} transmission response would also be exactly identical to Figure 2b. Thus, at operating frequency, the transmission magnitude is −6 dB for the AP_L value of 6.6 mm (the largest aperture dimension listed in Table 1), and the transmission magnitude reduces with the decreasing aperture values. This means that the loss of energy of the radiated wave occurs within the superstrate structure, and thus diminishes the radiation efficiency.

To validate the antenna's scattering performance, simulated monostatic RCS plotted against frequency for normal illumination of the incident wave is shown in Figure 8. Vertically polarized (VP) and horizontally polarized (HP) incident wave cases are respectively given in Figure 8a,b. Also shown in the figures is the RCS response against frequency of the reference antenna, which has the same lateral dimensions as of the proposed antenna. In addition, the calculated RCS frequency response of a perfect conductor of similar size is also presented. The calculation is based on the relation given as: $\sigma_C = 4\pi a^2/\lambda^2$, where σ_C is RCS of the perfect conductor, a is the area of the conductor, and λ represents wavelength of interest. For both polarizations shown in Figure 8a,b, wideband RCS reduction was achieved, including in-band frequencies. RCS reduction bandwidth (BW) extends over 4–16 GHz (120%), and an almost identical frequency response was achieved for the VP and HP wave incidences. This is owed to the symmetric unit cell design. For VP, average RCS reduction over the bandwidth is about 8.5 dB, and for HP, it is about 8.8 dB. Maximum achieved RCS reduction values are 25 dB (for VP) and 24 dB (for HP), respectively, appearing at 14.3 and 14.4 GHz value. The results shown in Figure 8a,b correspond to the co-polarized RCS frequency performance. The cross-coupled RCS monostatic performance of reference as well as proposed antenna, considering VP and HP wave incidences, is displayed in Figure 8c. As can be seen, the cross-coupled radar echoes are significantly low. The cross-coupled RCS performance of the proposed antenna obeys the cross-coupling reflection results presented for the unit cell in Figure 2a. Hence, the function of the absorber for reflectivity suppression is validated.

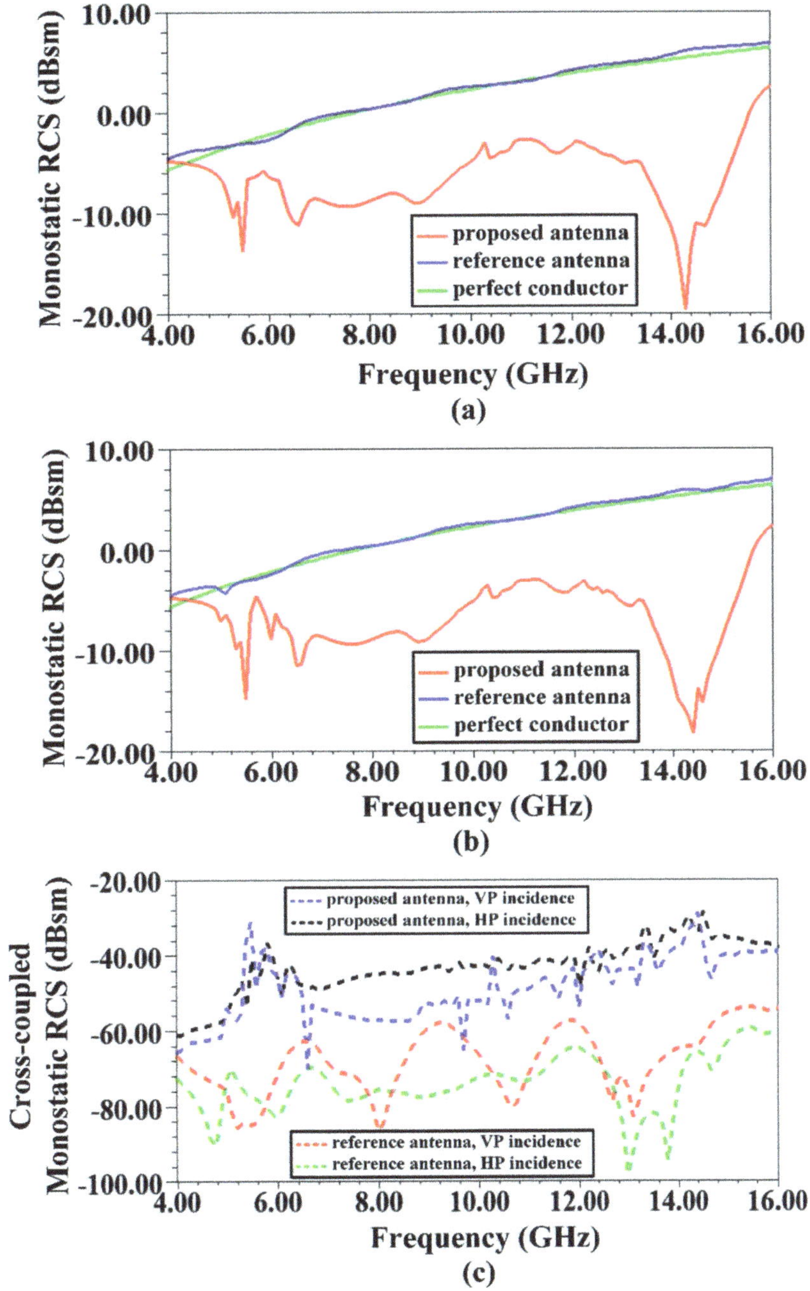

Figure 8. Frequency response of monostatic radar cross section (RCS) for (**a**) vertically polarized (VP) and (**b**) horizontally polarized (HP) wave incidences. (**c**) Cross-coupled monostatic RCS response for proposed and reference antenna.

The results presented in Figure 8 are for the case of 50 Ω antenna termination. This is because for most of the practical cases, the antenna would be matched terminated to its transmitter (Tx) or receiver (Rx). However, simulations for the cases of open/short terminations (worst case scenarios) were also performed for the proposed antenna, and it was observed that there was no significant change among different cases (short/open/matched). This might be due to the reason that the short or open primarily affects the antenna mode scattering, which is a function of antenna gain given as [41]:

$$\sigma_{ANT} = G^2\, \Gamma^2\, \lambda^2/4\pi, \tag{3}$$

where σ_{ANT} represents antenna mode scattering, G is antenna gain, Γ is reflection coefficient, and λ is the wavelength at the frequency of interest. The out-of-band frequencies will be unaffected by open/short terminations as these frequencies are being significantly absorbed at the absorber surface (and hence leave an insignificant energy that would reach the antenna surface). For the in-band frequencies, the in-band absorption is impaired (as evident from Figure 2a,b), but the in-band antenna mode scattering (frequencies reflected from the mismatched load and getting re-radiated by the antenna) would still be insignificant as the gain towards broadside is considerably low.

The measurement of frequency response of monostatic RCS was performed using two sets of in-lab developed parabolic reflector antennas that had lambda/2 dipole as their excitation source. Each set contained two such antennas (one for transmit (Tx) and one for receive (Rx) operation). Within the desired frequency range, six different frequencies were tested. In comparison to horn antennas, the size of these antennas is smaller; hence, in a real scenario, they can better replicate the ideal monostatic simulation setup—a merit that is worth noting. Figure 9a represents the simulated vs. measured monostatic RCS suppression for VP incident wave, whereas Figure 9b displays the same for the HP incident wave. The measurement setup is displayed in Figure 9c. As evident from Figure 9a,b, an agreeable correlation is found between the simulated and measured results; in fact, the measured suppression performance is superior for most of the frequency points. The details of tested frequencies with a comparison between simulation and measured RCS suppression levels are exhibited in Table 2. The measured mean suppression surpasses the simulated values. From these measurements, it can be asserted that the design verification stands as successful.

Table 2. RCS suppression values at tested frequencies.

Ser.	Tested Frequencies (GHz)	RCS Suppression (dB)			
		VP Incident Wave		HP Incident Wave	
		Simulation	Measurement	Simulation	Measurement
1.	7.5	9.32	11.49	9.44	12.92
2.	8	9.15	12.72	9.27	13.73
3.	8.8	9.93	18.48	10	18.3
4.	10	7.43	12.5	7.66	11.71
5.	12.6	8.61	13.16	9	14.51
6.	14	17.02	16.95	18.78	16
Mean RCS Suppression (dB)		10.24	14.22	10.69	14.53

The monostatic RCS was also investigated as a function of incidence angle (−90 to +90) to determine the angular stability of the proposed design. Figure 10a,b show normalized y–z plane plots for VP incident wave, respectively, for reference antenna and proposed antenna. Similarly, Figure 11a,b show normalized x–z plane plots for HP incident wave, respectively, for reference antenna and proposed antenna. Measured results of the above are presented along with, and depict, a satisfactory correlation with the simulation results. All of these results were plotted at 10 GHz. From Figures 10 and 11, RCS reduction occurs over about ±15° angular span. All presented RCS results are for antennas terminated with matched load.

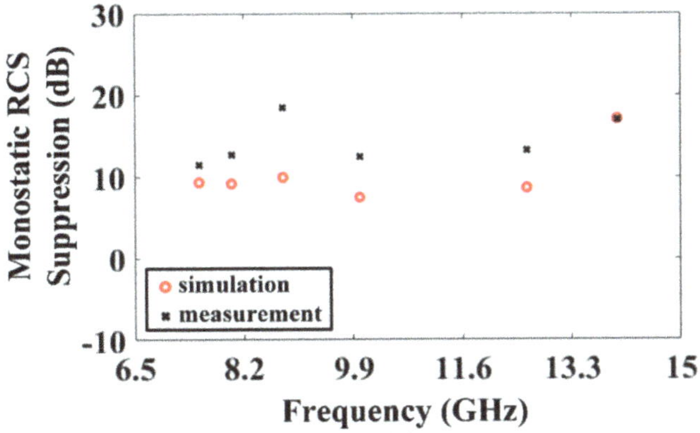

(a)

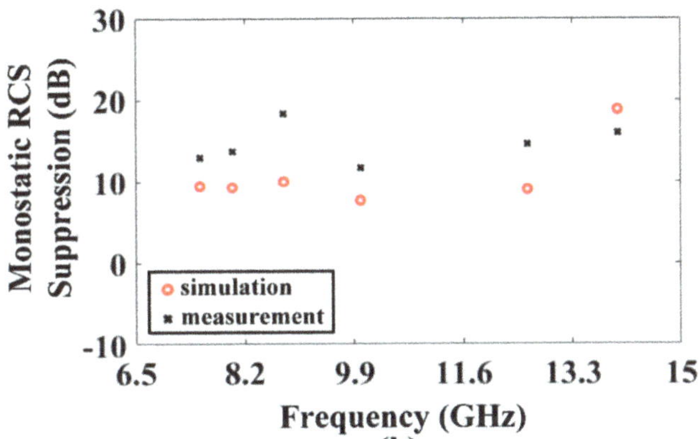

(b)

(c)

Figure 9. Simulation vs. measurement results of monostatic RCS suppression presented at different frequency points. (**a**) VP wave incidence. (**b**) HP wave incidence. (**c**) Measurement setup.

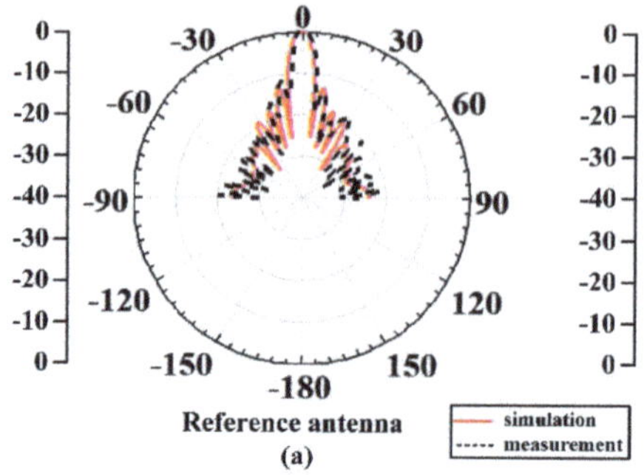

Figure 10. Normalized angular performance of monostatic RCS for y–z plane with VP incident wave at 10 GHz. (**a**) Reference antenna. (**b**) Proposed antenna.

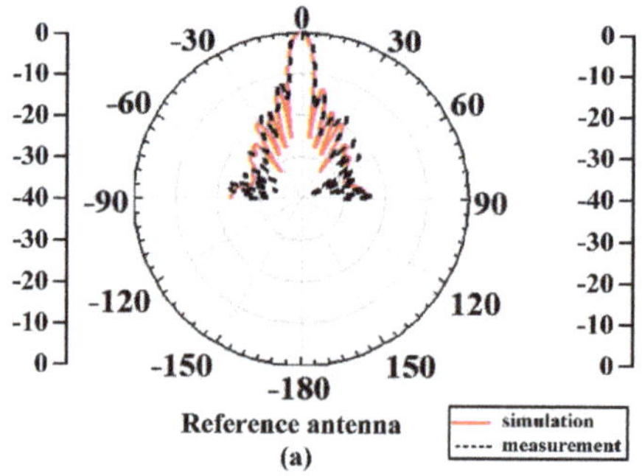

Figure 11. Normalized angular performance of monostatic RCS for x–z plane with HP incident wave at 10 GHz. (**a**) Reference antenna. (**b**) Proposed antenna.

In Figure 12, the simulated monostatic RCS angular responses of the reference and proposed antennas are presented for some more frequencies. Figure 12a illustrates the y–z plane RCS plots for a VP incident wave at 7.6 GHz, while Figure 12b illustrates the x–z plane RCS plots for an HP incident wave at 7.6 GHz. Similarly, Figure 12c illustrates the y–z plane RCS plots for a VP incident wave at 14 GHz, while Figure 12d illustrates the x–z plane RCS plots for an HP incident wave at 14 GHz. For the 7.6 GHz frequency, the RCS reduction in both planes occurs over an angular span of $\pm 22^\circ$, and it increases to $\pm 28^\circ$ at 14 GHz. It is inferred that as the broadside RCS reduction increases, the angular span over which the RCS reduction is achieved, increases as well. Furthermore, it can be observed from Figures 10–12 that in comparison to the reference antenna, the low RCS performance of the proposed antenna gets impaired towards wider off-broadside angles. In fact, the reflectivity even increases at too far-off angles. This increase may be attributed to the vertical dimension of the proposed antenna, as it is a Fabry–Perot cavity, while the reference antenna is a single layer planar antenna. Thus, it is important to mention that although the proposed antenna's main beam (radiation) is deflected towards off-broadside direction, the wideband low RCS performance of the proposed antenna is dominantly towards broadside angles only. Furthermore, as explained by the working of the unit cell in Section 2, the off-broadside peak radiation was achieved by implementing the phase gradated PRS, while the wideband broadside RCS reduction is a result of the PRS backed AFSS. Therefore, the antenna may potentially be used in those low observable military applications where it is desired to have communication in an off-broadside direction.

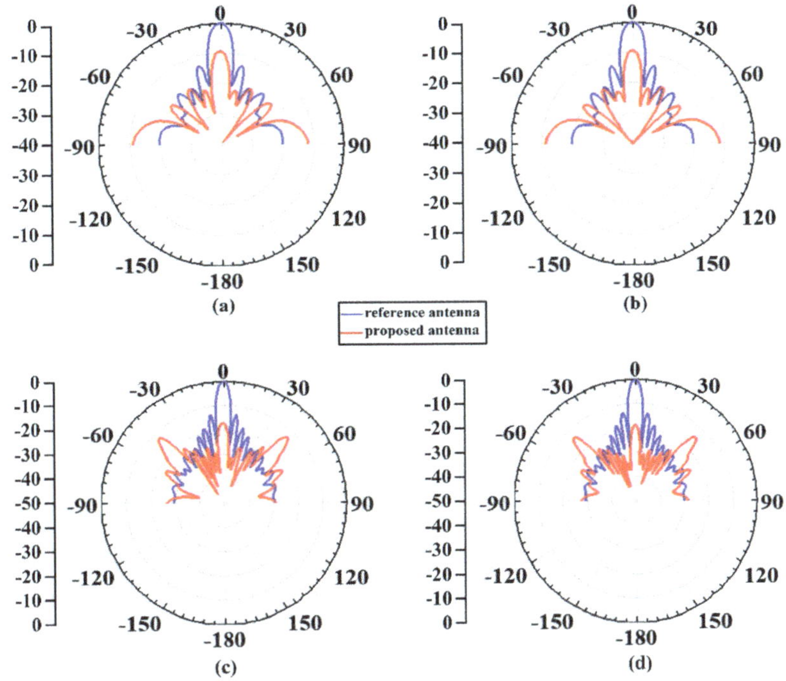

Figure 12. Normalized monostatic RCS angular response at various frequencies. (**a**) y–z plane, VP incident wave at 7.6 GHz. (**b**) x–z plane, HP incident wave at 7.6 GHz. (**c**) y–z plane, VP incident wave at 14 GHz. (**d**) x–z plane, HP incident wave at 14 GHz.

The monostatic RCS reduction for oblique angle incidence is shown in Figure 13. For the VP wave incidence, RCS reduction is shown in Figure 13a, and for the HP wave incidence, the RCS reduction is shown in Figure 13b. From the curves, it can be observed that for the incident angles of 5° and 10°, the performance is close to that of 0° case (boresight). For the wave incidence of 15°, the RCS reduction for both polarizations is close to 0 dB at the frequency points of 6, 10, and 11 GHz. It is important to mention that at these frequencies, both the proposed as well as the reference antennas realize nulls (or steep slopes just around the nulls) in their monostatic reflectivity patterns. For the incidence angle of 20°, the average RCS reduction becomes somewhat lower, and the RCS reduction value at 12 GHz is a little compromised. Thus, it can be asserted that the overall angular stability of the proposed design is slightly less than ±20°.

To have further insight into absorption as well as off-broadside radiation mechanism, surface E-field plots are presented in Figure 14. The log magnitude of E-field plotted on the absorber surface is show in Figure 14a. The plot is for a VP (x-polarized) incident wave, and at the frequency of 9 GHz. By inspection of Figure 14a, the spots of higher field intensity are easily identifiable. This field appears across the gaps in the metallic loops where resistors are present. The energy dissipation of the incident field occurs as heat within these resistors, thereby leading to the absorption of incident wave at the surface. Identical field plot is expected for an HP incident wave, except that the higher field intensity spots would now appear across the other two gaps for every loop. Likewise, field overlay plot on the surface containing gradated apertures (phase gradient surface) is shown in Figure 14b. This plot represents the field induced as a result of the wave radiated from the patch antenna (6 GHz frequency). It can be witnessed that a steady variation of surface E-field appears along the y-axis, in that, the E-field concentration increases gradually along –y-axis. This validates the mechanism behind beam deflection operation.

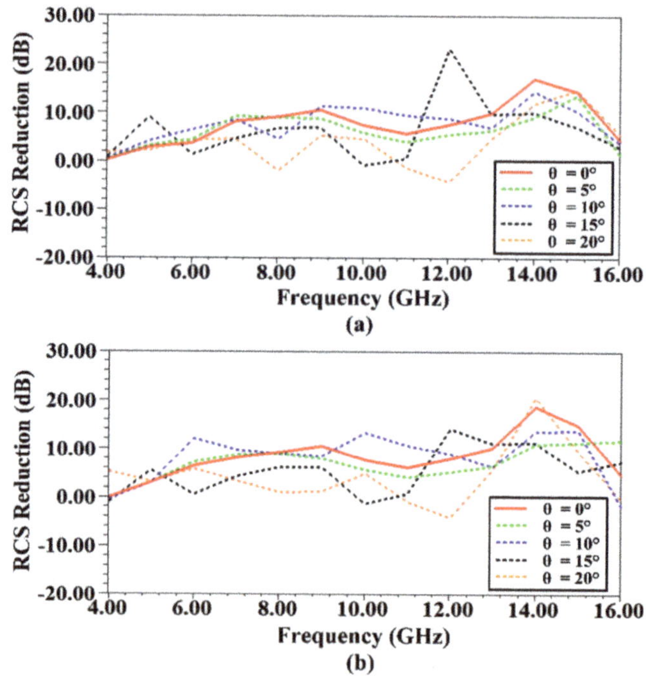

Figure 13. Simulated monostatic RCS reduction for oblique angle incidence of 5°, 10°, 15°, and 20°. (**a**) VP wave incidence. (**b**) HP wave incidence.

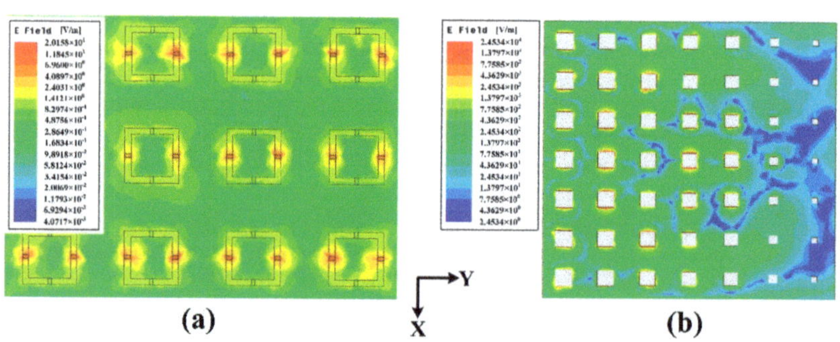

Figure 14. Surface E-field plots. (**a**) Absorber surface for VP incident wave. (**b**) Gradient surface.

4. Discussion

Table 3 shows a comparison of the proposed design with similar works from literature. In terms of RCS reduction, it is evident that the antenna performs almost equally well compared to most of the other designs. Additionally, the proposed antenna is capable of realizing an off-broadside beam radiation functionality. Furthermore, the achieved RCS reduction bandwidth is also superior to other works. In addition, the cavity height is also smaller, making it a low-profile design. Although proposed antenna also has high gain (relative to a conventional patch antenna), the value of gain quoted for other designs is higher as those antennas radiate a broadside beam, and hence do not suffer from scan loss [42]. The second reason for this is their aperture sizes, which are comparatively larger than the proposed design.

Table 3. Comparison of proposed work with previously reported designs.

Ref.	Size	Gain	Operating Frequency	RCS Reduction Band (% BW)	Average RCS Reduction	Beam Deflection
[25]	2.3 λ × 2.3 λ × 0.58 λ	13.2 dB	11.5 GHz	6–14 GHz (80%)	10 dB	No
[28]	2.79 λ × 2.79 λ × 0.68 λ	13.7 dB	9.05–10 GHz (reconfigurable)	7–14 GHz (66.67%)	unspecified	No
[29]	3.36 λ × 3.36 λ × 0.7 λ	17.9 dB	8.4 GHz	7–15 GHz (72.73%)	9.9 dB	No
[26]	3.7 λ × 3.7 λ × 0.73 λ	18.4 dB	10 GHz	8–17 GHz (72%)	13 dB	No
[43]	3.67 λ × 3.67 λ × 0.62 λ	19.8 dB	10 GHz	8–12 GHz (40%)	8.76 dB	No
This work	2.1 λ × 2.1 λ × 0.26 λ	9.4 dB	6 GHz	4–16 GHz (120%)	8.5 dB (VP) 8.8 dB (HP)	Yes (−38°)

5. Conclusions

A low-profile high gain FPC antenna that can simultaneously realize low backscattering as well as enhanced gain deflected beam operation has been presented in this article. To construct the cavity, an absorptive FSS was designed to work in conjunction with a double-sided PRS, and mounted on top of a patch radiator. One side of PRS is a capacitive grid, while the other side is an inductive grid. A dimensional gradient was implemented in the inductive part. For an incident wave, wideband RCS reduction was achieved and also included in-band frequencies. In the transmission mode of antenna, high gain as well as off-broadside beam radiation was achieved. Antenna cavity height is ~λ/4. The low scattering property of the proposed antenna makes it suitable to be integrated with stealth type platforms for communication, and it can find multiple applications in the military and defense realm; examples include side looking air borne radars, surveillance UAVs, and any military application that requires above/below the horizon communication. In future, the work can be extended to incorporate linear/circularly polarized MIMO antennas.

Author Contributions: Conceptualization, H.U. and T.B.A.L.; methodology, H.U. and T.H.; software, H.U. and T.H.; investigation, H.U.; resources, T.B.A.L., Y.Y., and K.K.; writing—original draft preparation, H.U.; writing—review and editing, T.B.A.L.; supervision, T.B.A.L., W.N.L.B.W.M., M.O., M.I.H.; project administration, T.B.A.L., Y.Y., W.N.L.B.W.M., and K.K.; funding acquisition, T.B.A.L., Y.Y., and M.I.H. All authors have read and agreed to the published version of the manuscript.

Funding: This research was funded by University of Malaya (UM), grant number GPF015A-2019.

Institutional Review Board Statement: Not applicable.

Informed Consent Statement: Not applicable.

Data Availability Statement: Not applicable.

Conflicts of Interest: The authors declare no conflict of interest.

References

1. Wiesbeck, W.; Heidrich, E. Influence of antennas on the radar cross section of camouflaged aircraft. In Proceedings of the 92 International Conference on Radar, Brighton, UK, 12–13 October 1992; pp. 122–125.
2. Wilsen, C.; Davidson, D. The radar cross section reduction of microstrip patches. In Proceedings of the IEEE AFRICON'96, Stellenbosch, South Africa, 27 September 1996; pp. 730–733.
3. Jia, Y.; Liu, Y.; Gong, S.-X.; Hong, T.; Yu, D. Printed UWB end-fire Vivaldi antenna with low RCS. *Prog. Electromagn. Res.* **2013**, *7*, 11–20. [CrossRef]
4. Jiang, W.; Liu, Y.; Gong, S.; Hong, T. Application of bionics in antenna radar cross section reduction. *IEEE Antennas Wirel. Propag. Lett.* **2009**, *8*, 1275–1278. [CrossRef]
5. Hu, S.; Chen, H.; Law, C.L.; Shen, Z.; Zhu, L.; Zhang, W.; Dou, W. Backscattering cross section of ultrawideband antennas. *IEEE Antennas Wirel. Propag. Lett.* **2007**, *6*, 70–73. [CrossRef]
6. Thakare, Y. Design of fractal patch antenna for size and radar cross-section reduction. *IET Microw. Antennas Propag.* **2010**, *4*, 175–181. [CrossRef]
7. Wang, W.; Gong, S.; Wang, X.; Guan, Y.; Jiang, W. Differential evolution algorithm and method of moments for the design of low-RCS antenna. *IEEE Antennas Wirel. Propag. Lett.* **2010**, *9*, 295–298. [CrossRef]

8. Yang, J.; Shen, Z. A thin and broadband absorber using double-square loops. *IEEE Antennas Wirel. Propag. Lett.* **2007**, *6*, 388–391. [CrossRef]
9. Miao, Z.; Huang, C.; Ma, X.; Pu, M.; Ma, X.; Zhao, Q.; Luo, X. Design of a patch antenna with dual-band radar cross-section reduction. *Microw. Opt. Technol. Lett.* **2012**, *54*, 2516–2520. [CrossRef]
10. Pozar, D.M. RCS reduction for a microstrip antenna using a normally biased ferrite substrate. *IEEE Microw. Guided Wave Lett.* **1992**, *2*, 196–198. [CrossRef]
11. Genovesi, S.; Costa, F.; Monorchio, A. Low-profile array with reduced radar cross section by using hybrid frequency selective surfaces. *IEEE Trans. Antennas Propag.* **2012**, *60*, 2327–2335. [CrossRef]
12. Costa, F.; Genovesi, S.; Monorchio, A. A frequency selective absorbing ground plane for low-RCS microstrip antenna arrays. *Prog. Electromagn. Res.* **2012**, *126*, 317–332. [CrossRef]
13. Costa, F.; Monorchio, A. A frequency selective radome with wideband absorbing properties. *IEEE Trans. Antennas Propag.* **2012**, *60*, 2740–2747. [CrossRef]
14. Chen, H.; Hou, X.; Deng, L. Design of frequency-selective surfaces radome for a planar slotted waveguide antenna. *IEEE Antennas Wirel. Propag. Lett.* **2009**, *8*, 1231–1233. [CrossRef]
15. Li, Y.-Q.; Zhang, H.; Fu, Y.-Q.; Yuan, N.-C. RCS reduction of ridged waveguide slot antenna array using EBG radar absorbing material. *IEEE Antennas Wirel. Propag. Lett.* **2008**, *7*, 473–476.
16. Zhang, J.; Wang, J.; Chen, M.; Zhang, Z. RCS reduction of patch array antenna by electromagnetic band-gap structure. *IEEE Antennas Wirel. Propag. Lett.* **2012**, *11*, 1048–1051. [CrossRef]
17. Gao, Q.; Yin, Y.; Yan, D.-B.; Yuan, N.-C. Application of metamaterials to ultra-thin radar-absorbing material design. *Electron. Lett.* **2005**, *41*, 936–937. [CrossRef]
18. Zheng, Y.; Gao, J.; Cao, X.; Yuan, Z.; Yang, H. Wideband RCS reduction of a microstrip antenna using artificial magnetic conductor structures. *IEEE Antennas Wirel. Propag. Lett.* **2015**, *14*, 1582–1585. [CrossRef]
19. Ren, J.; Gong, S.; Jiang, W. Low-RCS monopolar patch antenna based on a dual-ring metamaterial absorber. *IEEE Antennas Wirel. Propag. Lett.* **2017**, *17*, 102–105. [CrossRef]
20. Gao, J.; Zhao, Y.; Liu, T. A low RCS waveguide slot antenna array with metamaterial absorber. *IEEE Trans. Antennas Propag.* **2015**. [CrossRef]
21. Liu, Y.; Zhao, X. Perfect absorber metamaterial for designing low-RCS patch antenna. *IEEE Antennas Wirel. Propag. Lett.* **2014**, *13*, 1473–1476. [CrossRef]
22. Liu, T.; Cao, X.; Gao, J.; Zheng, Q.; Li, W.; Yang, H. RCS reduction of waveguide slot antenna with metamaterial absorber. *IEEE Trans. Antennas Propag.* **2012**, *61*, 1479–1484. [CrossRef]
23. Liu, Y.; Li, K.; Jia, Y.; Hao, Y.; Gong, S.; Guo, Y.J. Wideband RCS reduction of a slot array antenna using polarization conversion metasurfaces. *IEEE Trans. Antennas Propag.* **2015**, *64*, 326–331. [CrossRef]
24. Zheng, Q.; Guo, C.; Ding, J. Wideband and low RCS circularly polarized slot antenna based on polarization conversion of metasurface for satellite communication application. *Microw. Opt. Technol. Lett.* **2018**, *60*, 679–685. [CrossRef]
25. Pan, W.; Huang, C.; Chen, P.; Ma, X.; Hu, C.; Luo, X. A low-RCS and high-gain partially reflecting surface antenna. *IEEE Trans. Antennas Propag.* **2013**, *62*, 945–949. [CrossRef]
26. Jiang, H.; Xue, Z.; Li, W.; Ren, W.; Cao, M. Low-RCS high-gain partially reflecting surface antenna with metamaterial ground plane. *IEEE Trans. Antennas Propag.* **2016**, *64*, 4127–4132.
27. Mu, J.; Wang, H.; Wang, H.Q.; Huang, Y. Low-RCS and gain enhancement design of a novel partially reflecting and absorbing surface antenna. *IEEE Antennas Wirel. Propag. Lett.* **2017**, *16*, 1903–1906. [CrossRef]
28. Huang, C.; Pan, W.; Ma, X.; Luo, X. A frequency reconfigurable directive antenna with wideband low-RCS property. *IEEE Trans. Antennas Propag.* **2016**, *64*, 1173–1178. [CrossRef]
29. Jia, Y.; Liu, Y.; Zhang, W.; Wang, J.; Gong, S.; Liao, G. High-Gain Fabry-Perot Antennas with Wideband Low Monostatic RCS Using Phase Gradient Metasurface. *IEEE Access* **2018**, *7*, 4816–4824. [CrossRef]
30. Sumantyo, J.T.S. Design of tilted beam circularly polarized antenna for CP-SAR sensor onboard UAV. In Proceedings of the 2016 International Symposium on Antennas and Propagation (ISAP), Okinawa, Japan, 24–28 October 2016; pp. 658–659.
31. Lee, C.U.; Noh, G.; Ahn, B.; Yu, J.-W.; Lee, H.L. Tilted-Beam Switched Array Antenna for UAV Mounted Radar Applications with 360° Coverage. *Electronics* **2019**, *8*, 1240. [CrossRef]
32. Hassan, T.; Khan, M.U.; Attia, H.; Sharawi, M.S. An FSS Based Correlation Reduction Technique for MIMO Antennas. *IEEE Trans. Antennas Propag.* **2018**, *66*, 4900–4905. [CrossRef]
33. Thummaluru, S.R.; Kumar, R.; Chaudhary, R.K. Isolation enhancement and radar cross section reduction of MIMO antenna with frequency selective surface. *IEEE Trans. Antennas Propag.* **2018**, *66*, 1595–1600. [CrossRef]
34. Mehmood, A.; Karabey, O.H.; Jakoby, R. Dielectric resonator antenna with tilted beam. *IEEE Antennas Wirel. Propag. Lett.* **2016**, *16*, 1119–1122. [CrossRef]
35. Foroozesh, A.; Shafai, L. Investigation into the effects of the patch-type FSS superstrate on the high-gain cavity resonance antenna design. *IEEE Trans. Antennas Propag.* **2009**, *58*, 258–270. [CrossRef]
36. Costa, F.; Monorchio, A.; Manara, G. Analysis and design of ultra thin electromagnetic absorbers comprising resistively loaded high impedance surfaces. *IEEE Trans. Antennas Propag.* **2010**, *58*, 1551–1558. [CrossRef]

37. Nakano, H.; Mitsui, S.; Yamauchi, J. Tilted-beam high gain antenna system composed of a patch antenna and periodically arrayed loops. *IEEE Trans. Antennas Propag.* **2014**, *62*, 2917–2925. [CrossRef]
38. Ghasemi, A.; Burokur, S.N.; Dhouibi, A.; de Lustrac, A. High beam steering in Fabry–Pérot leaky-wave antennas. *IEEE Antennas Wirel. Propag. Lett.* **2013**, *12*, 261–264. [CrossRef]
39. Ourir, A.; Burokur, S.; de Lustrac, A. Phase-varying metamaterial for compact steerable directive antennas. *Electron. Lett.* **2007**, *43*, 493–494. [CrossRef]
40. Qin, F.; Gao, S.; Mao, C.; Wei, G.; Xu, J.; Li, J. Low-profile high-gain tilted-beam Fabry-Perot antenna. In Proceedings of the 2015 9th European Conference on Antennas and Propagation (EuCAP), Lisbon, Portugal, 13–17 April 2015; pp. 1–5.
41. Li, W.-Q.; Cao, X.-Y.; Gao, J.; Zhang, Z.; Cong, L.-L. Broadband RCS reduction and gain enhancement microstrip antenna using shared aperture artificial composite material based on quasi-fractal tree. *IET Microw. Antennas Propag.* **2016**, *10*, 370–377. [CrossRef]
42. Mailloux, R.J. *Phased Array Antenna Handbook*; Artech House: Norwood, MA, USA, 2017.
43. Zhang, L.; Wan, X.; Liu, S.; Yin, J.Y.; Zhang, Q.; Wu, H.T.; Cui, T.J. Realization of low scattering for a high-gain Fabry–Perot antenna using coding metasurface. *IEEE Trans. Antennas Propag.* **2017**, *65*, 3374–3383. [CrossRef]

applied sciences

MDPI

Article

A Refined Shape Sensing Method for Skin Antenna Structure Based on Inverse Finite Element Method

Shengtao Niu [1], Kexiang Li [2], Jianfeng Liu [1] and Hong Bao [1,*]

1 Key Laboratory of Electronic Equipment Structure Design of Ministry of Education, Xidian University, Xi'an 710071, China; stniu@stu.xidian.edu.cn (S.N.); liujf@xidian.edu.cn (J.L.)

2 724th Research Institute of China State Shipbuilding Corporation Limited, Nanjing 211100, China; lkxiang86@163.com

* Correspondence: hbao@xidian.edu.cn; Tel.: +86-029-8820-3040

Received: 3 August 2020; Accepted: 18 October 2020; Published: 29 October 2020

Abstract: An important issue in the existing inverse finite element method (iFEM) is that reconstruction accuracy cannot satisfy the analytical demand for the flexible structure. To address this issue, this paper presents a multi-nodes iFEM that reconstructs the displacement of structure based on surface measurement strains in real time. Meanwhile, in light of the response characteristics of iFEM, an innovative interpolation method is adapted to regenerate the full field deformation again. The proposed method substantially expands the size of inverse elements, which reduces the numbers of sensors and improves the reconstruction accuracy. The effectiveness of the method to predict displacement is verified by a flexible antenna panel subjected typical boundary conditions.

Keywords: reconstruction accuracy; multi-node iFEM; interpolation method; flexible antenna panel

1. Introduction

Today, with the rapid development of integrated antenna of radar technology, morphing-wing-embedded antenna arrays have an extensive application, such as military and civilian aerospace [1]. However, these structures are subjected to complex loads, such as wind and atmospheric pressure, which will produce various types of deformation leading to decrease the pointing accuracy of array antenna dramatically [2]. Therefore, real-time measurement of the deformation of antenna unit, which provides feedback to actuation and control systems of structure, is significant for human and environment safety [3,4].

The method on dynamically monitoring the deformation of structure, known as shape sensing, and the process of precisely approximating the integrated displacements based on strain measurement belongs to strictly the inverse problem [5]. To settle the inverse problem, Tikhonov et al. proposed a method included a regularization term to enhance the smoothness degree for approximating solution [6], and the technique was extensively employed for the inverse procedure. During the past few decades, the utilization of shape sensing algorithms for practical engineering has been demonstrated extensively. However, these approaches are mainly classified into the inverse finite element method (iFEM), modal method, and Ko's displacement method. Haugse and Foss proposed a modal method derived from the idea of polynomial basis function or spline function interpolation [7], which establishes the relationship between strain and deformation mode coordinates explicitly. Reference [8] use the modal method to reconstruct the displacement field of the beam structure precisely. However, the reconstruction accuracy is strictly limited on the number of sensors, which are at least higher than the mode orders extracted from structure. Thus, the method inevitably needs a high fidelity physical model, which is hard to do outside the laboratory environment. Then, Ko et al. develop a strategy, referred to as Ko's displacement method [9], which is based on the Euler–Bernoulli beam theory, and are extensively applicable to structural reconstruction such as beams, wing-boxes, and plates. The technique included the displacement superposition principle, which builds the quantitative relation about strain and element degrees of freedom definitely, resulting in the true shape sensing of slender structures. References [10,11] employ the numerical and experimental means to examine its feasibility. Although the method has a sufficiently accurate reconstruction of a slender structure, it is not suitable for the regeneration of three-dimensional structures. Hence, a well-suited shape-sensing algorithm, in addition to satisfying high accuracy, fast calculation, and strong robustness, also needs to consider the complexity of structural boundary conditions and geometric topology. Due to the deficiencies of the modal and the KO method, they are not suited for broad use in the field of structural health monitoring (SHM). Tessler and Spangler proposed an innovative methodology [12], named the inverse finite element method (iFEM), embodying the aforementioned characteristics required for a powerful SHM algorithm. The technology is based on the minimization of weighted-least-squares functional, which generally is applicable to complex structures subjected to complicated boundary conditions at any time [13]. The framework is sufficiently precise, powerful, and fast for applications in an engineering structure loaded statically and dynamically. Furthermore, the strain–displacement relationship was merely considered in the formulation, and materials of structure can be strictly ignored [14]. Benefiting from the potentiality of iFEM, the complex structure can be equivalent to simple models topologically, such as beam, plate, or shell. Moreover, based on the kinematic assumption of the Timoshenko beam theory, an inverse frame finite element is developed by Cerracchio and Gherlone [15]. References [16–19] have some analyses of three-dimensional frame structures under static or damped harmonic excitations based on strains produced numerically and experimentally. Based on the kinematic assumptions of the first-order shear deformation theory (FSDT), Tessler et al. developed three-node inverse shell element shorted form iMIN3 [20], which profiles the complex plate and shell structures accurately. Kefal et al. formulated a four-node quadrilateral inverse-shell element [21], which takes the influence of hierarchical drilling rotation into account, avoiding the occurrence of the singular value well. However, due to the heterogeneous and anisotropy of composite and sandwich structures, iFEM based on FSDT may result in slightly insufficient estimates for shape sense. Then, based on Zigzag displacement deformation theory, Cerracchio et al. and Kefal et al. improved the iFEM, which is suitable for sensing complex structures [22,23].

As mentioned above, the iFEM has prominent advantages, such as robustness, efficiency, and real-time response. However, the method needs numbers of inverse elements for shape sensing of flexible structures, leading to increasing the quantities of installing sensor. Therefore, the method is not directly suitable for a sophisticated antenna structure. This paper establishes a novel interpolation method to regenerate the structural deformation field based on data from the multi-node inverse finite element method. The technology optimally compromises the number of sensors and the inverse element and simultaneously retains the reconstruction accuracy.

The outline of the paper can be summarized as follows. First, the deformation field theory for plate is presented, and the multi-node iFEM formula is derived based on first-order shear deformation theory (FSDT). Second, taking the five-node quadrilateral inverse finite element as numerical implementation to result in the structure deformation. After that, the displacement field of the practical antenna panel is generated by the proposed method by experimental and numerical analysis approaches, respectively. According to some response dates from iFEM, the utilization of the innovative fitting technology restructured the full deformation field of board very well. The results from numerical and experimental analysis verify the correctness of the method proposed. Finally, the conclusions about the superiorities of the refined restructure methodology are emphasized.

2. Inverse Finite Element Formulation

Consider a plate with thickness of $2t$ as the analytical model, and the structure can be described by the general Cartesian coordinate system, as shown in Figure 1.

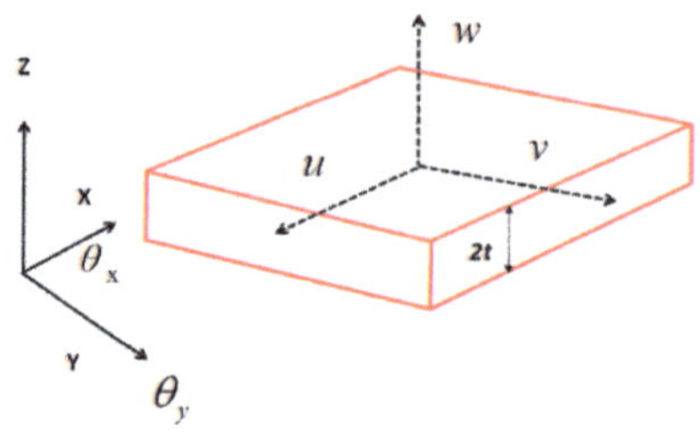

Figure 1. Plate model notion.

According to the Mindlin plate theory, the three-directional displacements relevant to arbitrary points can be expressed [13].

$$u_x(x, y, z) = u(x, y) + z\theta_y(x, y) \tag{1}$$

$$u_y(x, y, z) = v(x, y) + z\theta_x(x, y) \tag{2}$$

$$u_z(x, y, z) = w(x, y) \tag{3}$$

where, u_x, u_y are the plane displacements and u_z is the transverse displacement across the thickness orientation; θ_x, θ_y are rotations around the x- and y-axes, respectively; and z is the distance from the selected points to mid-plane of the structure.

The linear strain components are expressed by appropriate derivatives of the displacements.

$$\begin{aligned} E &= \{e_1, e_2, e_3\}^T = \left\{u_{x,x}, v_{y,y}, u_{x,y} + v_{y,x}\right\}^T \\ B &= \{e_4, e_5, e_6\}^T = \left\{\theta_{x,y}, \theta_{y,x}, \theta_{x,x} + \theta_{y,y}\right\}^T \\ G &= \{e_7, e_8\}^T = \left\{w_{z,x} + \theta_y, w_{z,y} + \theta_x\right\}^T \end{aligned} \tag{4}$$

where, E, B, G indicates membrane, bending curvatures, and transverse-shear strain measures, respectively.

According to the strain relation between any point and three main directions from material mechanics, the strain at any point of structure can be expressed.

$$\varepsilon(x_i, y_i, \theta) = \varepsilon_x \cos^2\theta + \varepsilon_y \sin^2\theta - \gamma_{xy} \sin\theta\cos\theta \tag{5}$$

Based on the functional theory, the Euclidean distance relationship between the measured strain and the theoretical strain can be expressed.

$$\begin{aligned} \phi(u) \;&= \sum_{i=1}^{n} w_i(\|\varepsilon(x_i, y_i, \theta_i) - \varepsilon^e(x_i, y_i, \theta_i)\|^2 + \|\varepsilon(x_i, z_i, \theta_i) - \varepsilon^e(x_i, z_i, \theta_i)\|^2 \\ &+ \|\varepsilon(y_i, z_i, \theta_i) - \varepsilon^e(y_i, z_i, \theta_i)\|^2) \end{aligned} \tag{6}$$

The ε^e is the measured strain from sensor, ε is the theoretical strains associated with location, which is pasted angle along axis (x, y, z), respectively, and n denotes the numbers of sensors, w_i is the weighting constants controlling the completeness between the analytic strains and their experimentally measured values.

Based on interpolation method, the deformation of points in the element can be expressed linearly by the shape function of element and the deformation of nodes. The result of any points in element can be expressed as follows:

$$\{u, v, w\} = \sum_{i=1}^{m} N_i u_i^e, (i = 1, 2, \ldots m) \tag{7}$$

where, N_i indicates the interpolation shape function of element, m is the numbers of element nodes, and u_i^e is the degree of freedom of element node expressed by nodal vector.

$$u_i^e = \begin{bmatrix} u_i & v_i & w_i & \theta_{xi} & \theta_{yi} & \theta_{zi} \end{bmatrix}^T, i = (1, 2, \ldots m) \tag{8}$$

Introducing Equations (4) and (5) and into Equation (6), and taking the minimum value of it about node degrees of freedom, gives rise to the function of shape sensing simultaneously.

$$k^e u^e = f^e \tag{9}$$

where, k^e is the element matrix related to only the location layout of strain sensors and the f^e vector is a function of the number of strain sensors in the element as well as the measured angle values. These can be explicitly written, respectively, as

$$k^e = w_i \sum_{i=1}^{n} (E^e + z \cdot B^e)^T m_i^T (E + z \cdot B^e) + r \iint (G^e)^T G^e dxdy \tag{10}$$

$$f^e = w_i \sum_{i=1}^{n} (E^e + z \cdot B^e)^T m_i^T \cdot \varepsilon_i^e \tag{11}$$

where, E^e, B^e, G^e can be listed by Appendix A, m_i is the angle cosine vector, and r is usually assigned a minimum value [13].

When the element matrix equations are established, the global deformation field can be performed explicitly according to the transformation relationship from the local to global coordinates.

$$KU = F \tag{12}$$

As K includes the rigid body motion mode of the discretized structure, therefore, it is a singular matrix. By combining problem-specific displacement boundary conditions, the resulting system of equations can be reduced from Equation (12)

$$K_a U_a = F_a \tag{13}$$

where, K_a is a 6m × 6m square matrix and invertible [13,24], F_a is a $6m \times 1$ matrix, and U_a is a $6m \times 1$ vector. The solution of Equation (12) is very efficient, because the matrix K_a remains unchanged for a determined configuration of sensors in the process of constant deformation. Moreover, F_a needs to be updated during any deformation cycle in real time. Finally, the unknown vector U_a reflects the structural state as well at any time.

According to general space interpolation methods, all the methods satisfy the estimation formula [25]

$$u(x, y) = \sum_{i=1}^{m} w_i(x, y) u_i \tag{14}$$

where $u(x, y)$ is the response value at the interpolated point in the structure, w_i indicates weighting function associated to the i-th sample point corresponding to the value u_i, and m represents the numbers of sampling points.

However, the inverse elements divided are evidently bigger than that fem, and the full field deformation obtained by Equation (12) has a distinct difference. The approximation approach is not suited for larger inverse elements. Based on the idea of constructing shape function shown [26], one can adapt a refined interpolation function that the nodes may be in the element rather than only in the boundary. The method employs the properties of shape function to improve the precision for a specific problem. The shape function can be described in the discrete vector

$$N_i = [N_i(x_1, y_1), N_i(x_2, y_2), \ldots\ldots, N_i(x_n, y_n)] \tag{15}$$

where N_i is $i-th$ node shape function, (x_n, y_n) is the coordinate in the element, n indicates the number of selected positions, and the arbitrarily point of the structure is expressed according to the interpolation definition

$$u(x, y) = N_1u_1 + N_2u_2 + \ldots + N_iu_i \tag{16}$$

The interpolation methods satisfy the four properties on shape function, and the geometric characteristics of the structure are described in some sense, overcoming the deficiency of little physical significances.

3. Numerical and Experimental Examples

The deformation of the antenna reflector is the main factor that affects pointing accuracy In order to examine the accuracy of iFEM for reconstructing the deformation of the antenna plate, an analysis between numerical and experiment on the skin antenna plate was presented.

A rectangular plate has a length of 0.560 m, a height of 0.202 m, and a uniform thickness of 3.2 mm. The plate has an elastic modulus of 1.0 GPa and a Poisson's ratio of 0.3 and is meshed 675 quadrilateral elements in analysis software; however, the structure was divided only two inverse quadrilateral elements (shown Figure 2). In Figure 2, the numbers in red indicate the element nodes, and the square letters in black denote the sampling points, respectively.

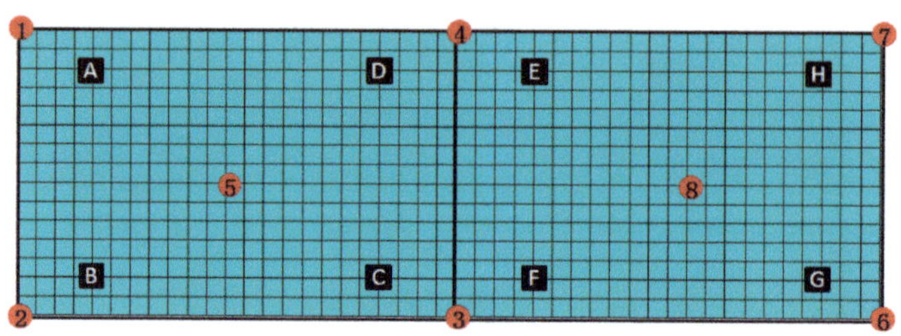

Figure 2. Inverse finite elements model of structure.

First, the deformation of sampling points can be obtained by iFEM. Second, based on these results, the whole deformation field can be reconstructed by the interpolation method proposed above. The deformations with two different boundary constraints are analyzed—namely, the left side of the structure—and both ends are fixed respectively.

With the condition of complete constraints at line connected node 1 and 2, the concentrated force of 10N is applied positive at nodes 5 and negative 10N at nodes 8 in the z direction concurrently. The nodes' deformations calculated by iFEM together with FEM are shown in Table 1.

Table 1. Results of nodes deformation.

Node	Z Directional Deformation (mm)		
	FEM	iFEM	Error
3	11.3	11.1	0.2
4	11.3	11.0	0.3
5	5.4	5.60	0.2
6	50.1	48.1	2.1
7	50.1	50.0	0.1
8	29.3	26.6	2.8

Based on the values of nodes 3–8 by iFEM, one can obtain the sample points A–H by the general interpolation method. Moreover, acting the nodes A–H together with 5 and 8 as sampling points, one can regenerate the full-field deformation with the proposed interpolation framework. The deformation of the whole structure from FEM and iFEM are presented in Figure 3.

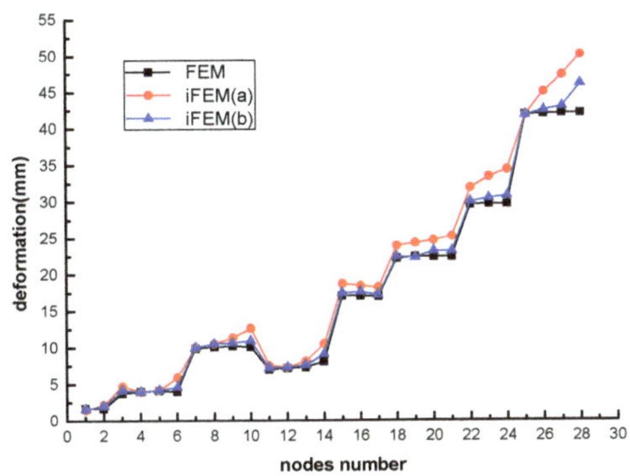

Figure 3. Deformation of the plate structure.

Figure 3 shows the deformed shapes of the plate obtained by different approaches shown from three different colors lines. The x-axis indicates the points of structure-expressed finite numbers, and y-axis is the deformation. According to the various interpolation means, iFEM(a) reflects the deformation trend of the whole structure according to the Equation (13), and iFEM(b) does that by Equation (15), respectively.

In order to quantitatively analyze the validity of the proposed method, the evaluation indexes mean error of nodes (*ME*) and root mean square (*RMS*) are introduced.

$$ME = \frac{1}{m}\sum_{i=1}^{m}\left|\left(X_i^{FEM} - Y_i^{iFEM}\right)\right| \quad (17)$$

$$RMS = \sqrt{\frac{\sum_{i=1}^{n}\left(X_i^{FEM} - Y_i^{iFEM}\right)^2}{n}} \quad (18)$$

where, X_i^{FEM} indicates the value provided by the finite element method, Y_i^{iFEM} is the value from the inverse finite element method, m and n are the number of nodes and sample points, respectively.

Considering the values from FEM as references, Table 1 lists the results from iFEM. The maximum deformation of the node along the z direction is 50.2 mm, and the maximum percentage difference is 8. The reconstruction *ME* is 1.87 mm, and the RMS of the reconstruction accuracy is 2.57 mm from iFEM(a). However, based on the proposed method, the RMS value of the deformation reconstruction accuracy is 1.0 mm.

With case two, lines connected nodes 1 and 2, and at the same time, 6 and 7 are fixed completely, so the concentrated force of 10N at node 5 and 10N at nodes 8 are applied in the positive z direction. In comparison, the deformation values of nodes are listed in Table 2. Based on these dates, the profile of structure can be reconstructed again by iFEM(b), as shown in Figure 4.

Table 2. Results of the nodes deformation.

Node	Z Directional Deformation (mm)		
	Fem	iFEM	Error
3	2.27	2.01	0.26
4	2.49	2.11	0.38
5	1.31	1.10	0.15
8	1.83	1.35	0.48

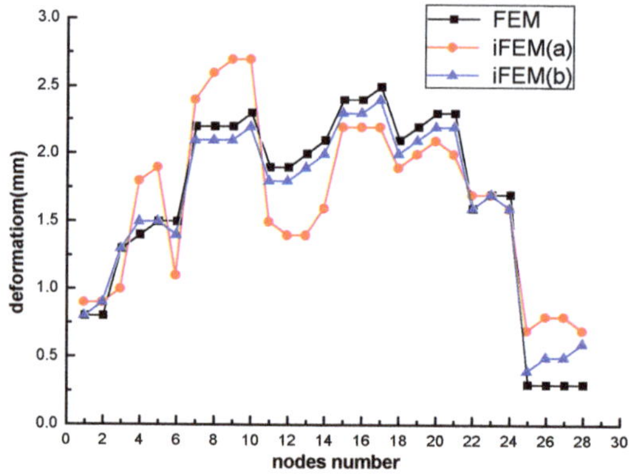

Figure 4. Deformation of the plate structure.

Table 2 displays the results from two approaches. The maximum deflection of the nodes is 2.5 mm, and the maximum deviation is 0.1 mm. The reconstruction *ME* is 0.3 mm, and the RMS of reconstruction accuracy is 0.3 mm. However, adopting the refined method, the RMS value of the deformation reconstruction accuracy decreases 0.1 mm.

Figures 3 and 4 depict the total deformed shapes of the plate employing different interpolation technologies iFEM(a) and iFEM(b) separately. For a view of profiles approaching FEM, the improved approximating skill raises the reconstruction accuracy dramatically.

In this section, one part of the antenna plate is considered. For the purpose of demonstrating the effectiveness put forward above, the panel has a length of 0.560 m, a height of 0.200 m, and a uniform thickness of 3.2 mm. For the aim of according with the simulated condition, 11 holes were arranged on the board, which was connected to fixed base by way of a bolt and screw. Meanwhile, the magnitude of the load is controlled by adjusting the nut up and down. The fixed device is shown (Figure 5a).

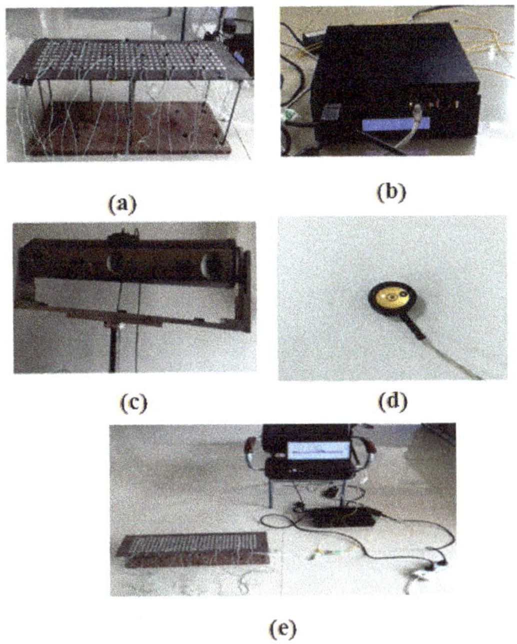

Figure 5. The measurement system; (**a**) fixed device; (**b**) Fiber Bragg Grating (FBG) demodulator; (**c**) Northern Digital Incorporation (NDI); (**d**) marker; (**e**) Measurement process.

According to the location sensors configuration from the simulation model, the fiber grating sensors are pasted at the corresponding positions of the experimental board one by one, and the markers with functions of photosensitization are pasted on the surface of the plate (Figure 5d). The variation of three-dimensional coordinates can be tracked by a 3D measurement system supplied by NDI (Waterloo, Ontario, Canada) (Figure 5c), and the surface strain date of structure can be read by the Fiber Bragg Grating (FBG) demodulator (Figure 5b).

Under the constraint conditions of the cantilever for the test article, the structure subjected to concentrated force is applied in the positive z direction. The inverse element is produced simplistically by these markers defined as element nodes at the edge. The markers number is the same as in Figure 2. The markers' initial and final position coordinates can be captured by NDI in real time, and the difference of the two sets of coordinates is the deformation of markers. Based on the measured strains and markers' variation, the results from the iFEM analysis and the NDI measurements are shown as Table 3, and the outline of the whole structure can be viewed in Figure 6.

Table 3. The results of markers deformation.

Marker	Z Directional Deformation (mm)		
	NDI	**iFEM**	**Error**
3	6.71	6.00	0.71
4	6.92	6.16	0.76
5	2.70	1.85	0.85
6	19.6	16.9	2.70
7	19.7	17.0	2.70
8	12.5	11.6	0.90

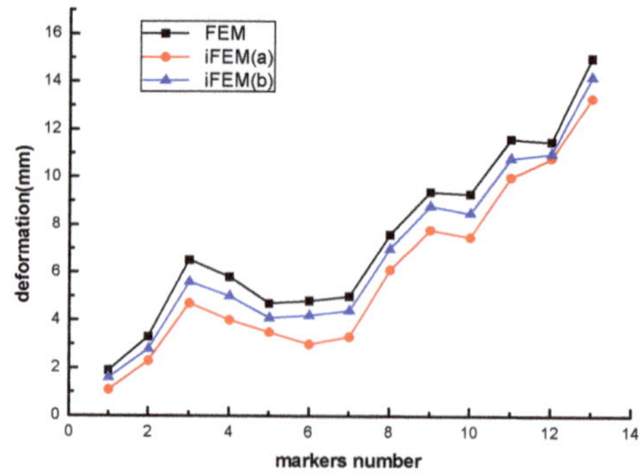

Figure 6. The deformation of the plate structure.

Table 3 lists the results generated from NDI measurements and iFEM, regarding the former as the benchmark. The maximum deformation of the structure in the Z direction is 19.7 mm. The reconstruction *ME* is 1.6 mm, and the RMS of reconstruction accuracy is 1.5 mm. Furthermore, on the basis of the proposed interpolation technology, the RMS value is 0.9 mm, and the reconstruction accuracy increased to 0.6 mm.

For the case of fixing two edges of structure, the results of markers' deformation are shown Table 4. In addition, the displacement of the full field can be achieved, as shown in Figure 7.

Table 4. The results of the markers deformation.

Marker	Z Directional Deformation (mm)		
	NDI	iFEM	Error
3	5.22	4.98	0.24
4	4.86	4.10	0.76
5	1.58	1.06	0.52
6	1.43	1.00	0.43

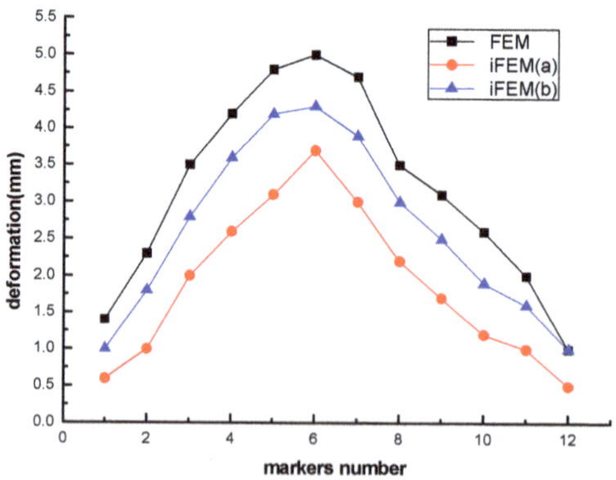

Figure 7. Deformation of plate structure.

The results from Table 4 state that the maximum deformation is 5.2 mm. The value of *ME* is 1.31 mm, and the RMS is 1.33 mm; otherwise, the RMS calculated by the proposed method is 0.80 mm. Figures 6 and 7 depict the deflection curve of integrated panel in different ways, considering the

NDI as an actual value, the curve smoothed by the improved interpolation method is preferable to approximate the real profile.

4. Discussion

Through the analysis of the estimated factors, the values significantly decrease in different degrees based on the proposed method. Meanwhile, the tendency of approaching the practicing deformation also verifies the effectiveness proposed method. Furthermore, the methodology is suitable for practical engineering applications.

5. Conclusions

A multi-node iFEM is presented for the shape sensing of a flexible structure with strain sensors installed at discrete locations. The method is based on the minimization of a weighted-least-squares functional that accounts for the complete set of strain measures consistent with Mindlin plate theory. Otherwise, based on the results from iFEM, adapting the refined interpolation technique to reconstruct the full field deformation again. The approach can have an accurate approximation for big inverse finite elements. Moreover, the method has the advantages of simultaneously achieving extrapolation and interpolation calculation for elements. Then, several case studies were performed and demonstrated the computational efficiency and high accuracy with respect to different forms of the loading structural responses.

The results demonstrated that sufficiently accurate reconstruction can be achieved for the problems considered herein based on a small number of sensors.

Author Contributions: S.N. carried out the majority of work on the paper, verifying measurement scheme, and writing the article. K.L. constructed the finite element model of antenna plate. H.B. proposed this new theory method for shape sensing of antenna plate in real time. J.L. gave advice on the writing of the article. All authors have read and agreed to the published version of the manuscript.

Funding: This research is financially supported by the National Natural Science Foundation of China through Grants 51775401, 51675398, and the CAS "Light of West China" Program 2016-QNXZ-A-7.

Conflicts of Interest: The author declares no conflict of interest.

Appendix A

The matrix E^e, B^e, G^e in Equations (9) and (10) can be defined as

$$E^e = \begin{bmatrix} N_{i,x} & 0 & 0 & 0 & 0 \\ 0 & N_{i,y} & 0 & 0 & 0 \\ N_{i,y} & N_{i,x} & 0 & 0 & 0 \end{bmatrix}$$

$$B^e = \begin{bmatrix} 0 & 0 & 0 & 0 & N_{i,x} \\ 0 & 0 & 0 & N_{i,y} & 0 \\ 0 & 0 & 0 & N_{i,y} & N_{i,x} \end{bmatrix}$$

$$G^E = \begin{bmatrix} 0 & 0 & N_{i,x} & 0 & 0 \\ 0 & 0 & N_{i,y} & 0 & 0 \end{bmatrix} (i = 1, 2, 3 \ldots m)$$

References

1. Kim, S.W.; Kim, E.H.; Rim, M.S.; Shrestha, P.; Lee, I.; Kwon, I.B. Structural Performance Tests of Down Scaled Composite Wind Turbine Blade using Embedded Fiber Bragg Grating Sensors. *Int. J. Aeronaut. Space Sci.* **2011**, *12*, 346–353. [CrossRef]
2. Arnold, E.J.; Yan, J.B.; Hale, R.D.; Rodriguez-Morales, F.; Gogineni, P. Identifying and Compensating for Phase Center Errors in Wing-mounted Phased Arrays for Ice Sheet Sounding. *IEEE Trans. Antennas Propag.* **2014**, *62*, 3416–3421. [CrossRef]

3. Akl, W.; Poh, S.; Baz, A. Wireless and distributed sensing of the shape of morphing structures. *Sens. Actuators A Phys.* **2007**, *140*, 94–102. [CrossRef]
4. Hopkins, M.A.; Truss, J.M.; Lockyer, A.J.; Alt, K.; Kinslow, R.; Kudva, J.N. Smart skin conformal load bearing antenna and other smart structures developments. In Proceedings of the 38th AIAA/ASME/ASCE/AHS/ASC, Structures, Structural Dynamics and Materials Conference, Kissimmee, FL, USA, 7–10 April 1997.
5. Maniatty, A.; Zabaras, N.; Stelson, K. Finite Element Analysis of Some Inverse Elasticity Problems. *J. Eng. Mech.* **1989**, *115*, 1303–1317. [CrossRef]
6. Maniatty, A.M.; Zabaras, N.J. Investigation of regularization parameters and error estimating in inverse elasticity problems. *Int. J. Numer. Methods Eng.* **1994**, *37*, 1039–1052. [CrossRef]
7. Foss, G.C.; Haugse, E.D. Using Modal Test Results to Develop Strain to Displacement Transformations. In Proceedings of the SPIE—The International Society for Optical Engineering, Philadelphia, PA, USA, 24–25 October 1995; Volume 2460, p. 112.
8. Bogert, P.B.; Haugse, E.D.; Gehrki, R.E. Structural shape identification from experimental strains using a modal transformation technique. In Proceedings of the 44th AIAA/ASME/ASCE/AHS Structures, Structural Dynamics and Materials Conference, Norfolk, VA, USA, 7–10 April 2003; p. 1626.
9. Ko, W.L.; Richards, W.L.; Fleischer, V.T. *Applications of the Ko Displacement Theory to the Deformed Shape Predictions of the Doubly-tapered Ikhana Wing*; NASA Technical Paper NASA/TP-2009-214652; NASA: Edwards, CA, USA, 2009.
10. Ko, W.L.; Fleischer, V.T. *Extension of KOStraight-beam Displacement Theory to Deformed Shape Predictions of Slender Curved Structures*; NASA: Edwards, CA, USA, 2011.
11. Jutte, C.V.; Ko, W.L.; Stephens, C.A.; Bakalyar, J.A.; Richards, W.L. *Deformed Shape Calculation of a Full-Scale Wing Using Fiber Optic Strain Data from a Ground Loads Test*; Rept, TP-215975; NASA Langley Research Center: Hampton, VA, USA, 2011.
12. Tessler, A.; Spangler, J.L. *A Variational Principal for Reconstruction of Elastic Deformation of Shear Deformable Plates and Shells*; NASA TM-2003-192445; NASA: Edwards, CA, USA, 2003.
13. Tessler, A.; Spangler, J.L. Inverse FEM for full-field reconstruction of elastic deformations in shear deformable plates and shells. In Proceedings of the 2nd European Workshop on Structural Health Monitoring, Munich, Germany, 7–9 July 2004.
14. Tessler, A.; Roy, R.; Esposito, M.; Surace, C.; Gherlone, M. Shape Sensing of Plate and Shell Structures Undergoing Large Displacements Using the Inverse Finite Element Method. *Shock. Vib.* **2018**, *2018*, 1–8. [CrossRef]
15. Gherlone, M. *Beam Inverse Finite Element Formulation*; LAQ Report; Politecnico di Torino: Turin, Italy, 2008.
16. Gherlone, M.; Cerracchio, P.; Mattone, M.; di Sciuva, M.; Tessler, A. Beam shape sensing using inverse Finite Element Method: Theory and experimental validation. In Proceedings of the 8th International Workshop on Structural Health Monitoring, Stanford, CA, USA, 13–15 September 2011.
17. Gherlone, M.; Cerracchio, P.; Mattone, M.; Di Sciuva, M.; Tessler, A. An inverse finite element method for beam shape sensing: Theoretical framework and experimental validation. *Smart Mater. Struct.* **2014**, *23*, 045027. [CrossRef]
18. Gherlone, M.; Cerracchio, P.; Mattone, M.; Di Sciuva, M.; Tessler, A. Dynamic shape reconstruction of three-dimensional frame structures using the inverse finite element method. In Proceedings of the 3rd ECCOMAS Thematic Conference on Computational Methods in Structural Dynamics and Earthquake Engineering, Island of Corfu, Greece, 26–28 May 2011.
19. Gherlone, M.; Cerracchio, P.; Mattone, M.; Di Sciuva, M.; Tessler, A. Shape sensing of 3D frame structures using an inverse Finite Element Method. *Int. J. Solids Struct.* **2012**, *49*, 3100–3112. [CrossRef]
20. Tessler, A.; Hughes, T.J.R. A three-node mindlin plate element with improved transverse shear. *Comput. Methods Appl. Mech. Eng.* **2012**, *50*, 71–101. [CrossRef]
21. Kefal, A.; Oterkus, E.; Tessler, A.; Spangler, J.L. A quadrilateral inverse-shell element with drilling degrees of freedom for shape sensing and structural health monitoring. *Eng. Sci. Technol. Int. J.* **2016**, *19*, 1299–1313. [CrossRef]
22. Cerracchio, P.; Gherlone, M.; di Sciuva, M.; Tessler, A. Shape and stress sensing of multilayered composite and sandwich structures using an inverse finite element method. In Proceedings of the 5th International Conference on Computational Methods for Coupled Problems in Science and Engineering, Ibiza, Spain, 17–19 June 2013.

23. Kefal, A.; Tessler, A.; Oterkus, E. An enhanced inverse Finite element method for displacement and stress monitoring of multilayered composite and sandwich structures. *Compos. Struct.* **2017**, *179*, 514–540. [CrossRef]
24. Barber, J.R. *Computational Elasticity: Theory of Elasticity, Finite and Boundary Element Methods*; Alpha Science Int'l Ltd.: Oxford, UK, 2008.
25. Baitsch, M.; Hartmann, D. Piecewise polynomial shape functions for hp-finite element methods. *Comput. Meth. Appl. Mech. Eng.* **2009**, *198*, 1126–1137. [CrossRef]
26. Shi, J.; Zheng, K.; Tan, Y.; Yang, K.; Zhou, G. Response simulating interpolation methods for expanding experimental data based on numerical shape functions. *Comput. Struct.* **2019**, *218*, 1–8. [CrossRef]

Publisher's Note: MDPI stays neutral with regard to jurisdictional claims in published maps and institutional affiliations.

applied sciences

MDPI

Article

Generation of Multiple Pseudo Bessel Beams with Accurately Controllable Propagation Directions and High Efficiency Using a Reflective Metasurface

Haixia Liu †, Hao Xue †, Yongjie Liu and Long Li *

Key Laboratory of High Speed Circuit Design and EMC of Ministry of Education, School of Electronic Engineering, Xidian University, Xi'an 710071, China; hxliu@xidian.edu.cn (H.L.); xuehao_xd@163.com (H.X.); liuyongjie_rf@163.com (Y.L.)

* Correspondence: lilong@mail.xidian.edu.cn

† These authors contributed equally to this work.

Received: 20 August 2020; Accepted: 9 October 2020; Published: 16 October 2020

Abstract: In this paper, a generation method procedure based on a reflective metasurface is proposed to generate multiple pseudo Bessel beams with accurately controllable propagation directions and high efficiency. Firstly, by adjusting the miniaturized unit cell of the reflective metasurface to modulate the electromagnetic waves using the proposed method, some off-axis pseudo Bessel beams with different propagation directions are generated. Then, by achieving the large-angle deflection and comparing the results with previous existing methods, the superiority of the proposed method is demonstrated. Based on the generated single off-axis pseudo Bessel beam and the superposition principle of the electromagnetic wave, a reflective metasurface with 47 × 47 elements is designed and fabricated at 10 GHz to generate dual pseudo Bessel beams. Full-wave simulation and experimental measurement results validate that the dual pseudo Bessel beams were generated successfully. The propagation directions of the dual pseudo Bessel beams can be controlled accurately by the reflective metasurface, and the efficiency of the beams is 59.2% at a propagation distance of 400 mm. The energy of the beams keeps concentrating along the propagation axes, which provides a new choice for wireless power transfer and wireless communication with one source to multiple receiving targets.

Keywords: off-axis pseudo Bessel beam; multiple pseudo Bessel beams; reflective metasurface; accurately controllable propagation direction; high efficiency

1. Introduction

Electromagnetic waves can realize the wireless transmission of energy and information, which provide an irreplaceable role for the development of society. Meanwhile, the divergence characteristic of an electromagnetic wave limits its application, and the Bessel beam [1] with non-diffracting characteristic shows the potential to solve this problem. Therefore, devices that can generate high-performance Bessel beams have been extensively studied, and a source with limited spatial size can only generate a quasi-Bessel beam, which has the ideal Bessel beam's characteristics over a finite non-diffracting area [2–5]. As the energy in the non-diffraction area has the characteristics of uniform and high efficiency, it is suitable for Bessel beams to be used in wireless power transfer (WPT) [6–8] and near-field communication systems [9,10]. The high-order Bessel beams with the characteristics that vortex beams have also show great potential in the field of high-speed wireless communication and high-resolution imaging [11–13]. However, most devices can only generate on-axis Bessel beams, and there are few studies on off-axis Bessel beams [14–21]. So, we studied some methods about steering Bessel beams or focused beams [22–29], and it is easy to find that, although the Bessel beams have a wider range of energy distribution, there are very few methods to steer them. One method is proposed to steer

Bessel beams using computer holograms with the help of amplitude modulation [22]. Another novel method can generate arbitrarily directed Bessel beams using an off-axis multi-focusing antenna [24], and this method is also helpful to improve the design precision of the depth-of-field of the beam. As the beam deflection requires only phase modulation, the lack of amplitude modulation will result in a deflected beam that is not strictly a Bessel beam, and we can call it a pseudo Bessel beam. It is not easy to maintain high efficiency and high directionality when the beam is deflected, and because the performance of the single off-axis pseudo Bessel beam is not satisfactory, multiple pseudo Bessel beams are more difficult to be achieved. This also leads to the fact that the devices generate Bessel beams that are not multifunctional, which makes them unable to give full play to their advantages in wireless transmission systems.

In this paper, we propose a method to generate a single off-axis pseudo Bessel beam and multiple pseudo Bessel beams with accurate propagation directions and high efficiency. To modulate electromagnetic waves effectively [30–32], we firstly use a reflective metasurface that has low loss to generate the off-axis pseudo Bessel beam in any desired direction. As the reflective metasurface is single-layer, it is also low profiled compared with non-planar or multi-layer devices [5,6]. The unit cell of the metasurface has a periodicity of 8 mm at 10 GHz, and it also has the characteristics of a small size, high efficiency, and insensitive to both polarization and incidence angle. As the method we proposed satisfies the requirement of a transverse wave vector component of the Bessel beam, comparing with existing methods, the generated off-axis pseudo Bessel beam using our method has an accurately controllable propagation direction and high efficiency. Then, according to the proposed method and electromagnetic field superposition principle, a metasurface with 47 × 47 unit cells is proposed to generate multiple pseudo Bessel beams simultaneously for the first time. The metasurface is simulated with a standard X-band horn in ANSYS HFSS and the results show that the proposed method can be used to generate dual pseudo Bessel beams. The generated beams have a transmission efficiency of 59.2% at a transmission distance of 400 mm and maintain high efficiency in the non-diffracting area. Finally, the metasurface is fabricated and measured, and the experimental measurement results show that our designed metasurface can effectively generate multiple pseudo Bessel beams. The proposed method and device have shown their effectiveness, which provide a new way for the application of one-to-many WPT or wireless communication.

2. Design Method of the Off-Axis Pseudo Bessel Beam and Multiple Pseudo Bessel Beams

A metasurface has a controllable phase shift to realize the control of the electromagnetic wave by adjusting the unit cells. As shown in Figure 1a, the metasurface on the xoy-plane can be used to generate pseudo Bessel beams, propagating in the directions of u_k or u_l, which means any direction in the coordinate system. It is important to calculate the compensation phase of each unit cell whose coordinate is $(x_{ij}, y_{ij}, 0)$, and the unit cell we proposed has the structure of two rings and one cross, which will be explained in detail in the next section. The compensation phase that the unit cell provides is controlled by the size of the unit cell, and the compensation phase distribution of the metasurface is decided by the beams we prepare to generate. As the generation of a pseudo Bessel beam needs the coherent superimposition of both inward conical waves and outward conical waves [13,33], it is necessary to modulate these waves by the metasurface. As shown in Figure 1a, the energy radiated from the feed horn to the metasurface can be modulated by the unit cells to form the inward conical wave. The inward conical wave is shown as dashed lines in Figure 1a,b; it is worth pointing out that all of the unit cells contribute for the inward conical wave so that they can form a conical surface. After passing through the propagation axis, the inward conical wave becomes the outward conical wave, and a pseudo Bessel beam will be formed in the overlapping area of the two. The wave vector k of the inward conical wave can be divided into the transverse component k_t and the longitudinal component k_l, i.e., $k = \sqrt{k_t^2 + k_l^2} = 2\pi/\lambda$ (λ is the wavelength in free space). It is easy to find that if the transverse components of the inward and outward conical waves have the same magnitude and opposite directions, a standing wave will be formed in the transverse direction. In other words, the angle α between the inward conical wave and the propagation axis ($\tan\alpha = k_t/k_l$) remains the same in this condition, and this will prohibit the energy to leak out in the transverse direction. Meanwhile, the longitudinal components of the inward and outward conical waves will have the same magnitude and direction, so a traveling wave can be formed in the longitudinal direction. Therefore, in the superimposed region, a pseudo Bessel beam can propagate along the propagation axis with all of the electromagnetic energy.

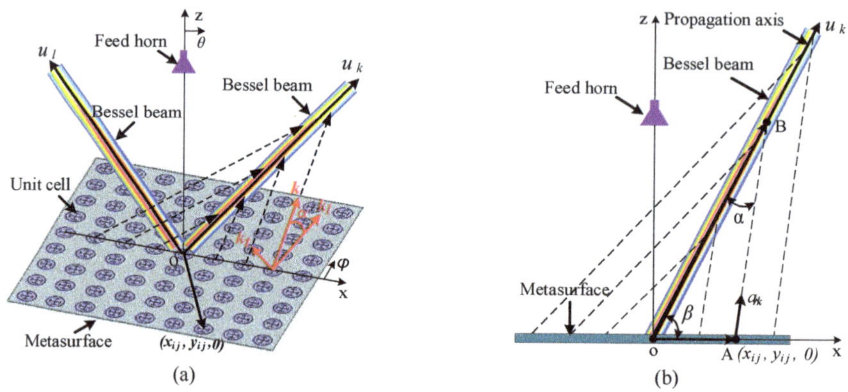

Figure 1. Schematic diagram of (**a**) pseudo Bessel beams generated by a reflective metasurface; (**b**) the proposed method to generate the off-axis pseudo Bessel beam.

From the mentioned generation procedure of a pseudo Bessel beam, we can know that forming a standing wave in the transverse direction is the key that all of the energy can propagate along the desired direction. This also means the included angle α between the inward conical wave and the propagation axis needs to remain the same, and as the focused beams do not meet this requirement, they cannot remain uniform in energy distribution during a long distance [26–29]. Only in this way can the "standing wave" condition be satisfied, and all of the energy can be propagated along the propagation axis without leaking to other directions. Based on the above principle and discussion, we can give a method for generating an off-axis pseudo Bessel beam using a metasurface.

As it is shown in Figure 1b, the first thing we need to do is to determine the propagation direction of the pseudo Bessel beam; the unit vector of the pseudo Bessel beam can be expressed as $\hat{u}_k = (\sin\theta_k\cos\varphi_k, \sin\theta_k\sin\varphi_k, \cos\theta_k)$, where θ_k and φ_k are the pitch angle and the azimuth angle of the desired beam in the coordinate system, respectively. Then we need to determine the ratio of the wave vector's transverse component to the longitudinal component; that is, determine the angle α. Based on these settings, the required compensation phase distribution on the metasurface can be calculated according to the condition of the transverse component of the wave vector and geometric optics. This method can satisfy the condition of forming a standing wave on the plane that is perpendicular to the beam propagation direction, thereby ensuring that all energy radiates along the propagation axis.

To get the compensation phase distribution of the reflective metasurface, which can generate the energy-focused, off-axis pseudo Bessel beam, the phase of the E-field at each point on the metasurface can be expressed as

$$\Phi_1(x_{ij}, y_{ij}) = -\frac{2\pi}{\lambda}\hat{a}_k\cdot\vec{OA} \tag{1}$$

where $\hat{a}_k$ is the unit vector along to the inward conical wave of each point, as shown in Figure 1b, and the vector corresponding to each point on the metasurface can be expressed as

$$\vec{OA} = \sqrt{x_{ij}{}^2 + y_{ij}{}^2}(\cos\varphi_{ij},\ \sin\varphi_{ij}, 0) \tag{2}$$

where $\varphi_{ij} = \arctan(y_{ij}/x_{ij})$. The vector corresponding to the intersection of the straight line in the direction of any vector $\hat{a}_k$ and the propagation axis is

$$\vec{OB} = \frac{\sin(180° - \beta - \alpha)}{\sin\alpha}\sqrt{x_{ij}{}^2 + y_{ij}{}^2}\hat{u}_k \tag{3}$$

where β is the included angle between the vectors $\vec{OA}$ and $\vec{OB}$, and it can be obtained by the vector operation

$$\cos\beta = \sin\theta_k\cos\varphi_k\cos\varphi_{ij} + \sin\theta_k\sin\varphi_k\sin\varphi_{ij} \tag{4}$$

Finally, the unit vector $\hat{a}_k$ can be expressed as

$$\hat{a}_k = \frac{(\vec{OB} - \vec{OA})}{\left|\vec{OB} - \vec{OA}\right|} \tag{5}$$

Taking Equation (5) into Equation (1), the phase $\Phi_1(x_{ij}, y_{ij})$ at each point of the metasurface can be calculated for the off-axis pseudo Bessel beam, which is also the compensation phase of each unit cell to transform the plane wave to the inward conical wave. However, the feed horn we actually use can only radiate the spherical wave. Therefore, we also need to calculate the compensation phase on the metasurface to convert the spherical wave radiated from the feed horn into the plane wave. The required compensation phase is

$$\Phi_2(x_{ij}, y_{ij}) = \frac{2\pi}{\lambda}(\sqrt{x_{ij}{}^2 + y_{ij}{}^2 + r_f{}^2} - r_f) \tag{6}$$

where r_f is the distance from the feed horn to the metasurface. Then, the final compensation phase of any off-axis pseudo Bessel beam generated by the feed horn and the metasurface is

$$\Phi_k(x_{ij}, y_{ij}) = \Phi_1(x_{ij}, y_{ij}) + \Phi_2(x_{ij}, y_{ij}) \tag{7}$$

In this way, we can get the compensation phase of each unit cell on the metasurface to generate the k-th off-axis pseudo Bessel beam, and the compensation phase to generate multiple pseudo Bessel beams can be gotten by using the electromagnetic field superposition principle. The phase compensation of each unit cell whose coordinate is $(x_{ij}, y_{ij}, 0)$ can be calculated by the following equation [34]:

$$\Phi_{total}(x_{ij}, y_{ij}) = \arg\{\sum_{k=1}^{n}\exp(j\Phi_k(x_{ij}, y_{ij}))\} \tag{8}$$

where j is the imaginary unit, Φ_k is the compensation phase that needed to be adjusted on the k-th beam, and n is the total number of the multiple pseudo Bessel beams.

3. Metasurface Simulation and Experimental Measurement Results

3.1. Metasurface Simulation and Results

As the method of calculating the compensation phase for generating a single off-axis pseudo Bessel beam and multiple pseudo Bessel beams by a metasurface has been proposed, we can use the suitable unit cell to compose the needed metasurfaces. A miniaturized subwavelength unit cell with a periodicity P = 8 mm (0.27λ) [13] is selected as the unit cell of the metasurface, as show in Figure 2. The unit cell comprises two concentric circular rings, while the inner ring is connected to a cross, and is also separated by four identical gaps with an angle of 20°. The radiuses of the outer ring and the inner ring are R_o and R_i, which has the relationship of $R_i = 0.7 \times R_o$, and the width of both the rings and the cross is w = 0.3 mm. This unit cell is printed on a F4B ($\varepsilon_r = 2.65$) dielectric substrate with a thickness of t = 3 mm, and the model of it is built in the simulation software ANSYS HFSS, with periodic boundary conditions. The reflection phase and reflection coefficient characteristics of the unit cell versus the length of R_o at the center frequency of 10 GHz are shown in Figure 2.

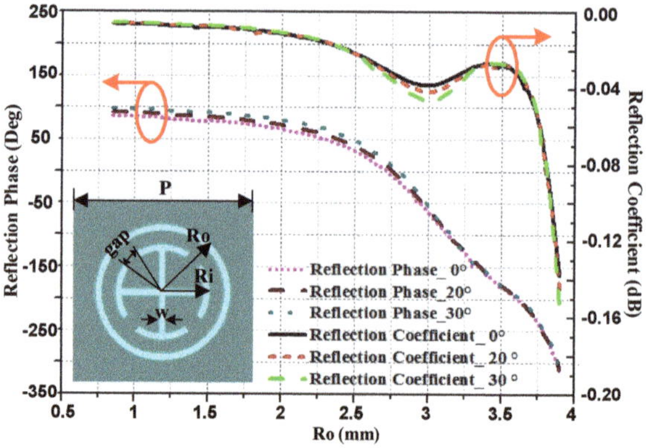

Figure 2. Geometry (top view) of the designed unit cell and the reflection phase and reflection coefficient versus the length of R_o in an infinite periodic model with different incidence angles.

As shown in Figure 2, by varying R_o from 0.85 to 3.9 mm, the reflection phase variation range of the unit cell can cover the whole 360°, and it can be seen that the phase-shift changes are less than 17° when the incidence angle varies from 0° to 30°. Meanwhile, the reflection coefficient keeps above −0.2 dB when the incidence angle varies, so that this miniaturized unit cell has the characteristics of both high efficiency and good angle stability; it is also polarization-insensitive because of its symmetrical structure.

According to the method mentioned and the relationship between the unit cell size and its reflection phase curves shown in Figure 2, the metasurfaces to generate off-axis pseudo Bessel beams can be designed. All of the metasurfaces have a size of 390 mm × 390 mm, and the distance from the feed horn to the metasurfaces is 300 mm. The azimuth angle of the beam is set to be $\varphi_k = 45°$, and the angle between the inward conical wave and the propagation axis is $\alpha = 10°$. In order to show the universality of this method, we chose the pitch angles to be 15°, 30°, and 45°, respectively. The designed metasurfaces were simulated by ANSYS HFSS, and the models and simulated results of the generated off-axis pseudo Bessel beams are shown in Figure 3.

The normalized E-field amplitude distributions in Figure 3 show that the off-axis pseudo Bessel beams, which have the predetermined orientations, were generated successfully. As the energy is concentrated along the propagation axis without leakage in other directions, the method we use shows its effectiveness. To demonstrate the superiority of the proposed method, we can compare it with the traditional method, which realizes the off-axis pseudo Bessel beam by adding the compensation phases of the on-axis Bessel beam and the deflected beam. As the differences between these two methods will become more obvious when the pitch angle becomes larger, we set the pitch angle to be 60°, and the other settings are just the same as in Figure 3. The normalized E-field amplitude distributions and corresponding compensation phase distributions of the off-axis pseudo Bessel beams using the two different methods are shown in Figure 4.

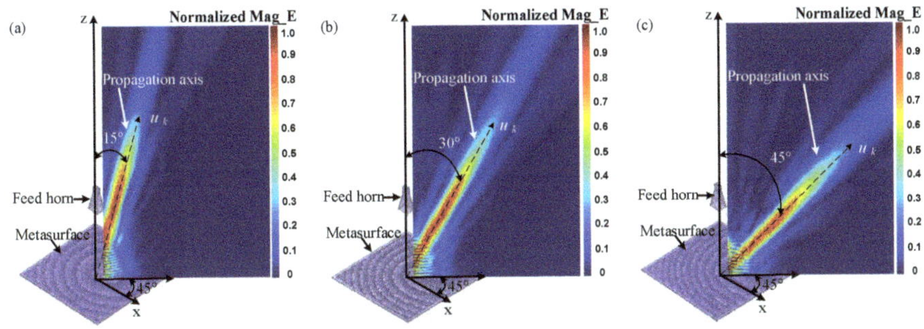

Figure 3. The normalized E-field amplitude distributions of the off-axis pseudo Bessel beams with the propagation directions of (**a**) $\theta_k = 15°$, $\varphi_k = 45°$; (**b**) $\theta_k = 30°$, $\varphi_k = 45°$; (**c**) $\theta_k = 45°$, $\varphi_k = 45°$.

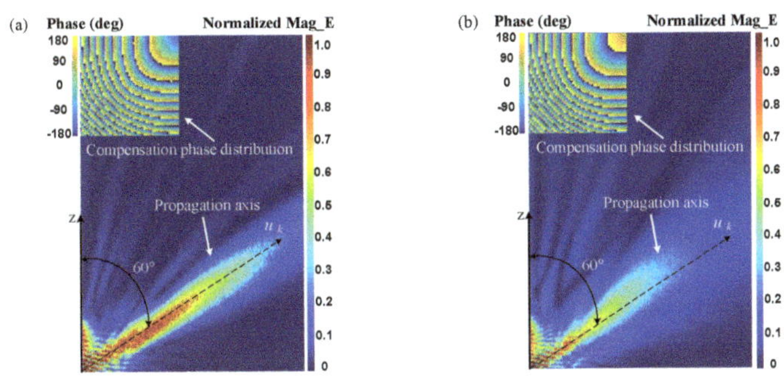

Figure 4. The normalized E-field amplitude distributions of the off-axis pseudo Bessel beams and their corresponding compensation phase distributions when using (**a**) the proposed method; (**b**) the traditional method.

It is easy to find that the compensation phase distributions of the two methods have the same trend, but the difference is also obvious. The simulated results show that the pseudo Bessel beam using the proposed method can keep the energy concentrating on the propagation axis in the non-diffraction area, while the divergence of energy is obvious in the traditional method. So compared with previous methods, the proposed method shows its superiority.

Based on the high-performance, off-axis pseudo Bessel beams we have generated, and the superposition of different off-axis pseudo Bessel beams according to Equation (8), multiple pseudo Bessel beams can also be generated simultaneously using the metasurface. We chose the dual pseudo Bessel beams as an example, and the propagation directions were set to be $\theta_1 = 30°$, $\varphi_1 = 0°$ and $\theta_2 = 30°$, $\varphi_2 = 180°$, respectively, while the angle $\alpha = 10°$ for both of the two beams. The unit cell was still chosen to be the unit cell shown in Figure 2, and the metasurface has the whole dimension of 390 mm × 390 mm × 3 mm, while the coordinate of the feed horn in the coordinate system was still 0 mm, 0 mm, and 300 mm. Then, the required compensation phase distribution, the layout of the metasurface, and the simulated normalized E-field amplitude distribution of the generated dual pseudo Bessel beams on the xoz-plane are shown in Figure 5.

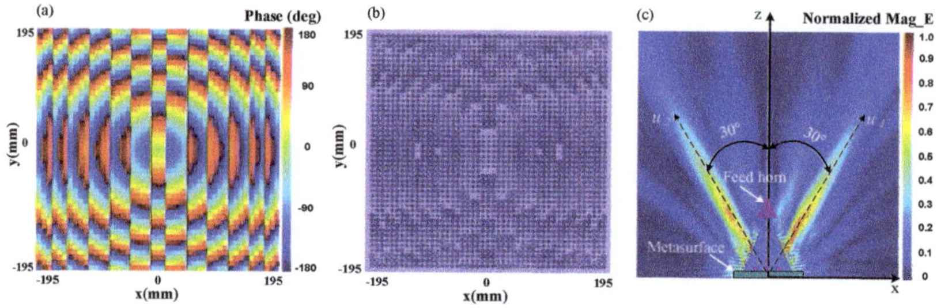

Figure 5. (**a**) The compensation phase distribution of the designed metasurface; (**b**) the layout of the metasurface; (**c**) the normalized E-field amplitude distribution of the dual pseudo Bessel beams on the xoz-plane.

It can be seen from Figure 5c that the dual pseudo Bessel beams have been generated by the metasurface simultaneously, and the propagation directions of them meet the predetermined settings. The energy is also concentrated along the propagation axes, and to further prove this, we can observe the normalized E-field amplitude distributions on two sets of observation planes that are, respectively, perpendicular to the two propagation directions. These observation planes with a size of 0.3 m × 0.3 m are placed at different distances from the metasurface, and the distances are 400 mm, 600 mm, 800 mm, 1000 mm, and 1200 mm, respectively.

As shown in Figure 6, it is obvious that the energy is concentrated on the center point of each observation plane, which is also the intersection of the propagation axis and the observation plane. In addition, the normalized E-field amplitudes of both the two beams decrease as the distances from the metasurface increase, while the normalized E-field amplitude distributions that on the same distance of the two beams are symmetrical. This phenomenon shows that the energy allocated to the two beams is basically the same, which is in line with our design.

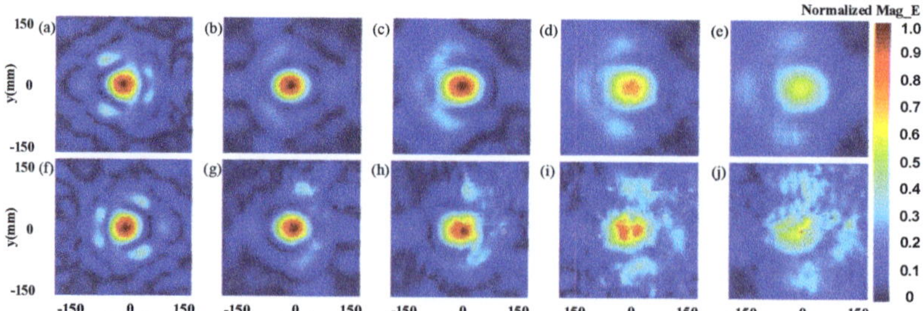

Figure 6. Simulated normalized E-field amplitude distributions on the different observation planes at the distances of (**a**) 400 mm, (**b**) 600 mm, (**c**) 800 mm, (**d**) 1000 mm, and (**e**) 1200 mm away from the metasurface when the azimuth angle of the pseudo Bessel beam is 0°. The simulated normalized E-field amplitude distributions on the different observation planes at the distances of (**f**) 400 mm, (**g**) 600 mm, (**h**) 800 mm, (**i**) 1000 mm, and (**j**) 1200 mm away from the metasurface when the azimuth angle of the pseudo Bessel beam is 180°.

3.2. Metasurface Experimental Measurement and Results

The fabricated metasurface is fed by a broadband horn antenna that operates from 2 to 18 GHz and is measured by using the near-filed planar-scanning techniques in the anechoic chamber, as shown in Figure 7a. And the prototype of the fabricated reflective metasurface is shown in Figure 7b.

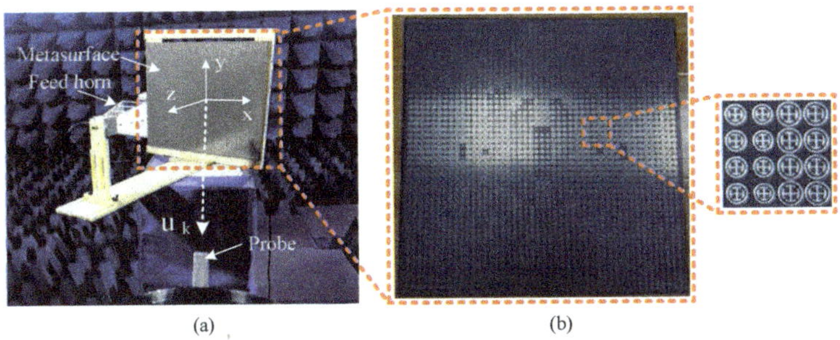

Figure 7. (**a**) Experimental measurement scene of the dual pseudo Bessel beams in the anechoic chamber; (**b**) the prototype of the fabricated reflective metasurface.

A standard waveguide probe is used to detect the normalized E-field amplitude distributions corresponding to that shown in Figure 6. To detect the off-axis beams, the metasurface should have an inclined angle of ±30° to the *z*-axis, and the scanning plane has a size of 0.3 m × 0.3 m with a sampling grid of 5 mm. The normalized E-field amplitude distributions on the different observation planes are, respectively, perpendicular to the two propagation directions and have a distance from the metasurface of 400 mm, 600 mm, 800 mm, 1000 mm, and 1200 mm, measured as shown in Figure 8. The comparison between Figures 6 and 8 shows that the experimental measurement results of the dual pseudo Bessel beams are in good agreement with the simulated one.

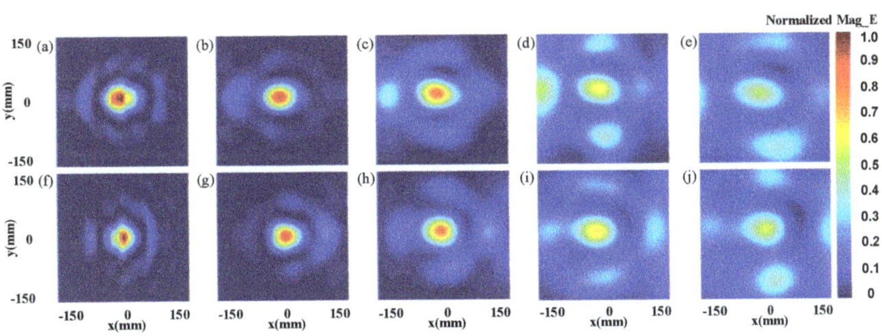

Figure 8. Experimental measurement normalized E-field amplitude distributions on the different observation planes at a distance of (**a**) 400 mm, (**b**) 600 mm, (**c**) 800 mm, (**d**) 1000 mm, and (**e**) 1200 mm away from the metasurface when the azimuth angle of the pseudo Bessel beam is 0°. The experimental measurement normalized E-field amplitude distributions on the different observation planes at a distance of (**f**) 400 mm, (**g**) 600 mm, (**h**) 800 mm, (**i**) 1000 mm, and (**j**) 1200 mm away from the metasurface when the azimuth angle of the pseudo Bessel beam is 180°.

In order to show the phenomenon of dual pseudo Bessel beams, we observe the normalized E-field amplitude distributions of the dual-beams on a 1.2 m × 0.6 m observation plane, which is perpendicular to the probe. The distance between the probe and the observation plane is 0.6 m, and the experimental measurement normalized E-field amplitude distribution on the observation plane is shown in Figure 9.

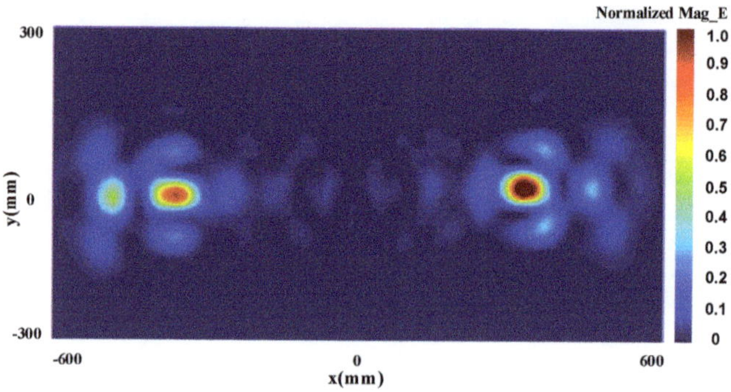

Figure 9. Experimental measurement normalized E-field amplitude distribution of the dual pseudo Bessel beams on the observation plane with a size of 1.2 m × 0.6 m and a distance to the probe of 0.6 m.

From the experimental measurement result shown in Figure 9, it is sure that the energy concentrates on the points where the pseudo Bessel beams pass through, and the amplitudes of the two beams are basically equivalent. So, we can draw the conclusion that the dual pseudo Bessel beams were generated by the metasurface, and almost all of the energy is propagating along the desired directions.

4. Discussion

The dual pseudo Bessel beams were generated using the designed metasurface with the proposed method. From the simulated normalized E-field amplitude distribution, we can see two pseudo Bessel beams propagate along the directions of $\theta_1 = 30°$, $\varphi_1 = 0°$ and $\theta_2 = 30°$, $\varphi_2 = 180°$ simultaneously, and the energy is divided into two almost equal parts along the desired directions. The energy along the propagation axes is nearly remaining stable, while the simulated and experimental measurement normalized E-field amplitudes at the different distances have the tendency of decreasing gradually, which means the energy is attenuating when propagating. We can calculate the transmission efficiency of the dual pseudo Bessel beams generated by the metasurface at different transmission distances to further research the energy distribution of the dual pseudo Bessel beams. As shown in Figure 10, the transmission efficiency (η) of the metasurface can be defined as the ratio of the pseudo Bessel beam power (P_1) to the total power (P_0) radiated by the metasurface, i.e., $\eta = P_1/P_0$, and the values of both P_0 and P_1 are the flux integral of the Poynting vector.

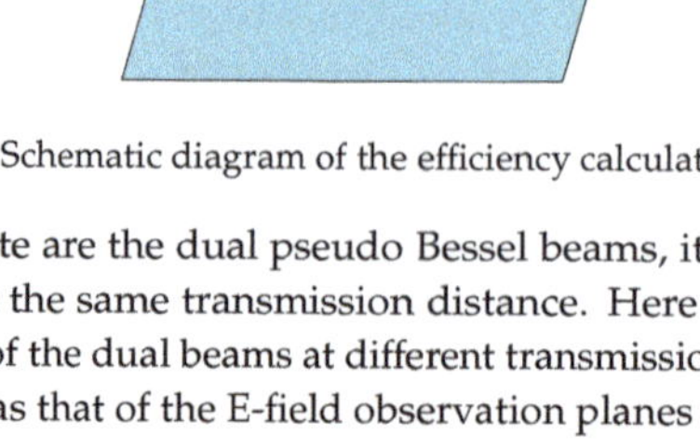

Figure 10. Schematic diagram of the efficiency calculation method.

As the beams we generate are the dual pseudo Bessel beams, it is necessary to add the power of the two beams obtained at the same transmission distance. Here we use five different reference planes to calculate the power of the dual beams at different transmission distances. The distances from the metasurface are the same as that of the E-field observation planes mentioned, so the transmission distances are 400 mm, 600 mm, 800 mm, 1000 mm, and 1200 mm, respectively. The result shows that on the reference plane has a distance of 400 mm, and the total calculated power value of the two beams is 0.474. As the power value P_0 on the metasurface can be obtained as 0.8 by integration, so the efficiency of the dual pseudo Bessel beams at 400 mm is about 59.2%. The efficiency at the latter distances was also calculated by the same method as being 36.8%, 32.6%, 27%, and 25%, respectively. It can be seen that the efficiency changes with the change in transmission distance, and it maintains a relatively high value considering that the reference planes are far away from the metasurface; that is, a high efficiency can be obtained at different transmission distances in accurately controllable propagating directions when using the proposed method to generate pseudo Bessel beams. As the energy can be allocated in multiple directions, devices at multiple different locations can be charged simultaneously. Furthermore, because the generated pseudo Bessel beams have uniform energy distributions within a certain range, the location requirements for the devices are also reduced.

5. Conclusions

In this paper, multiple pseudo Bessel beams can be effectively generated based on the method of generating an off-axis pseudo Bessel beam and the principle of superposition, and the beams have the advantages of both accurately controllable propagating directions and high efficiency. The metasurface for generating dual pseudo Bessel beams was designed, fabricated, and measured, and the simulated and experimentally measured normalized E-field distributions demonstrate that nearly all the radiated energy from the metasurface propagates along the designed directions. The full wave simulation results show that the dual beams have a high efficiency of 59.2% at a transmission distance of 400 mm from the metasurface, and the efficiency keeps relatively high in the non-diffracting area. The proposed method and device for generating multiple pseudo Bessel beams can be applied to multi-user WPT systems, wireless communication systems, and multi-objects radio frequency identification (RFID) systems with a large operating distance.

Author Contributions: H.L. and H.X. conceived the simulations and finished the paper writing; Y.L. analyzed the metasurface and performed the experiments; L.L. provided the main instructions of the ideas. All authors have read and agreed to the published version of the manuscript.

Funding: This work is supported by the Science and Technology Project of State Grid Corporation of China "Research on Ten-Meter Level Microwave Radio Wireless Power Transmission Technology".

Conflicts of Interest: The authors declare no conflict of interest.

References

1. Durnin, J. Exact solutions for nondiffracting beams. I. The scalar theory. *J. Opt. Soc.* **1987**, *4*, 651–654. [CrossRef]
2. Durnin, J.; Miceli, J., Jr.; Eberly, J.H. Diffraction-free beams. *Phys. Rev. Lett.* **1987**, *58*, 1499–1501. [CrossRef] [PubMed]
3. Ettorre, M.; Pavone, S.C.; Casaletti, M.; Albani, M. Experimental Validation of Bessel Beam Generation Using an Inward Hankel Aperture Distribution. *IEEE Trans. Antennas Propag.* **2015**, *63*, 2539–2544. [CrossRef]
4. Ettorre, M.; Rudolph, S.M.; Grbic, A. Generation of Propagating Bessel Beams Using Leaky-Wave Modes. *IEEE Trans. Antennas Propag.* **2012**, *60*, 3605–3613. [CrossRef]
5. Monk, S.; Arlt, J.; Robertson, D.A.; Courtial, J.; Padgett, M.J. The generation of Bessel beams at millimetre-wave frequencies by use of an axicon. *Opt. Commun.* **1999**, *170*, 213–215. [CrossRef]
6. Li, Y.B.; Cai, B.G.; Wan, X.; Cui, T.J. Diffraction-free surface waves by metasurfaces. *Opt. Lett.* **2014**, *39*, 5888–5891. [CrossRef]
7. Mazzinghi, A.; Balma, M.; Devona, D.; Guarnieri, G.; Mauriello, G.; Albani, M.; Freni, A. Large depth of field pseudo-Bessel beam generation with a RLSA antenna. *IEEE Trans. Antenna Propag.* **2014**, *62*, 3911–3919. [CrossRef]
8. Rodriguez, E.S.G.; Machnoor, M.; Lazzi, G. On the generation of nondiffracting beams in extremely subwavelength applications. *IEEE Trans. Antennas Propag.* **2017**, *65*, 5228–5237. [CrossRef]

9. Heebl, J.D.; Ettorre, M.; Grbic, A. Wireless links in the radiative near field via Bessel beams. *Phys. Rev. Appl.* **2016**, *6*, 034018. [CrossRef]
10. Imani, M.F.; Grbic, A. Generating Evanescent Bessel Beams Using Near-Field Plates. *IEEE Trans. Antennas Propag.* **2012**, *60*, 3155–3164. [CrossRef]
11. Kou, N.; Yu, S.; Li, L. Generation of high-order Bessel vortex beam carrying orbital angular momentum using multilayer amplitude-phase-modulated surfaces in radiofrequency domain. *Appl. Phys. Express* **2016**, *10*, 016701. [CrossRef]
12. Pfeiffer, C.; Grbic, A. Controlling Vector Bessel Beams with Metasurfaces. *Phys. Rev. Appl.* **2014**, *2*, 044012. [CrossRef]
13. Liu, H.; Xue, H.; Liu, Y.; Feng, Q.; Li, L. Generation of high-order Bessel orbital angular momentum vortex beam using a single-layer reflective metasurface. *IEEE Access* **2020**, *8*, 126504–126510. [CrossRef]
14. Lemaitre-Auger, P.; Abielmona, S.; Caloz, C. Generation of Bessel Beams by Two-Dimensional Antenna Arrays Using Sub-Sampled Distributions. *IEEE Trans. Antennas Propag.* **2013**, *61*, 1838–1849. [CrossRef]
15. Mei, Q.Q.; Wen, X.T.; Cui, T.J. A Broadband Bessel Beam Launcher Using Metamaterial Lens. *Sci. Rep.* **2015**, *5*, 11732.
16. Salem, M.A.; Kamel, A.H.; Niver, E. Microwave Bessel Beams Generation Using Guided Modes. *IEEE Trans. Antennas Propag.* **2011**, *59*, 2241–2247. [CrossRef]
17. Trappe, N.; Mahon, R.; Lanigan, W.; Murphy, J.A.; Withington, S. The quasi-optical analysis of Bessel beams in the far infrared. *Infrared Phys. Technol.* **2005**, *46*, 233–247. [CrossRef]
18. Vasara, A.; Friberg, A.T.; Turunen, J. Holographic generation of diffraction-free beams. *Appl. Opt.* **1988**, *27*, 3959.
19. Yi, C.Z.; Yu, J.C. Ka-Band Wideband Large Depth-of-Field Beam Generation through a Phase Shifting Surface Antenna. *IEEE Trans. Antennas Propag.* **2016**, *64*, 5038–5045.
20. Zhaofeng, L.; Boratay, A.K.; Humeyra, C.; Ekmel, O. Generation of an axially asymmetric Bessel-like beam from a metallic subwavelength aperture. *Phys. Rev. Lett.* **2009**, *102*, 143901.
21. Wei, T.C.; Khorasaninejad, M.; Zhu, A.Y.; Oh, J.; Devlin, R.C.; Zaidi, A.; Capasso, F. Generation of wavelength-independent subwavelength Bessel beams using metasurfaces. *Light Sci. Appl.* **2016**, *6*, e16259.
22. Salo, J.; Meltaus, J.; Noponen, E.; Westerholm, J.; Salomaa, M.M.; Lonnvist, A.; Saily, J.; Hakli, J.; Ala-Laurinaho, J.; Raisanen, A.V. Millimetre-wave Bessel beams using computer holograms. *Electron. Lett.* **2002**, *37*, 834–835. [CrossRef]
23. Yi, C.Z.; Yu, J.C. Wideband Quasi-Non-Diffraction Beam with Accurately Controllable Propagating Angle and Depth-of-Field. *IEEE Trans. Antennas Propag.* **2017**, *65*, 5035–5042.
24. Cheng, Y.; Zhong, Y.; Renbo, H.E.; Liu, Y.; Fan, Y.; Song, K.; Zhang, Y. Antenna for Generating Arbitrarily Directed Bessel Beam. U.S. Patent No. 10,644,398, 5 May 2020.
25. Wu, Y.; Cheng, Y. Generating and 2-D Steering Large Depth-of-Field Beam by Leaky-Wave Antenna Array with a Modified Parabolic Reflector. *IEEE Trans. Antennas Propag.* **2019**, *68*, 2779–2787. [CrossRef]
26. Wu, Y.; Cheng, Y.; Huang, Z. Ka-band near-field-focused 2-D steering antenna array with a focused Rotman lens. IEEE Transactions on Antennas and Propagation. *IEEE Trans. Antennas Propag.* **2018**, *66*, 5204–5213. [CrossRef]
27. Buffi, A.; Nepa, P.; Manara, G. Design criteria for near-field-focused planar arrays. *IEEE Antennas Propag. Mag.* **2012**, *54*, 40–50. [CrossRef]
28. Álvarez, J.; Ayestarán, R.G.; León, G.; Herrán, L.F.; Arboleya, A.; López-Fernández, J.A.; Las-Heras, F. Near field multifocusing on antenna arrays via non-convex optimisation. *IET Microw. Antennas Propag.* **2014**, *8*, 754–764. [CrossRef]
29. Ettorre, M.; Pavone, S.C.; Casaletti, M.; Albani, M.; Mazzinghi, A.; Freni, A. Near-field focusing by non-diffracting Bessel beams. In *Aperture Antennas for Millimeter and Sub-Millimeter Wave Applications, Signals and Communication Technology*; Springer: Rennes, France, 2018; Volume 8, pp. 243–288.
30. Shaltout, A.M.; Kildishev, A.V.; Shalaev, V.M. Evolution of photonic metasurfaces: From static to dynamic. *J. Opt. Soc. Am. B* **2016**, *33*, 501–510. [CrossRef]
31. Li, L.; Liu, H.; Zhang, H.; Xue, W. Efficient wireless power transfer system integrating with metasurface for biological applications. *IEEE Trans. Ind. Electron.* **2018**, *65*, 3230–3239. [CrossRef]
32. Li, L.; Zhang, X.; Song, C.; Huang, Y. Progress, challenges, and perspective on metasurfaces for ambient radio frequency energy harvesting. *Appl. Phys. Lett.* **2020**, *116*, 060501. [CrossRef]

33. Chávez-Cerda, S. A new approach to Bessel beams. *J. Mod. Opt.* **1999**, *46*, 923–930. [CrossRef]
34. Yu, S.; Li, L.; Shi, G.; Zhu, C.; Shi, Y. Generating multiple orbital angular momentum vortex beams using a metasurface in radio frequency domain. *Appl. Phys. Lett.* **2016**, *108*, 241901. [CrossRef]

MDPI
St. Alban-Anlage 66
4052 Basel
Switzerland
www.mdpi.com

Applied Sciences Editorial Office
E-mail: applsci@mdpi.com
www.mdpi.com/journal/applsci

www.ingramcontent.com/pod-product-compliance
Lightning Source LLC
Chambersburg PA
CBHW061623210926
51850CB00009B/2413